Digital Video Editing

A USER'S GUIDE

D0266676

CORNWALL COLLEGE

Digital Video Editing

A USER'S GUIDE

Peter Wells

THE CROWOOD PRESS

First published in 2007 by
The Crowood Press Ltd
Ramsbury, Marlborough
Wiltshire SN8 2HR

www.crowood.com

© Peter Wells 2007

All rights reserved. No part of this publication may be reproduced or transmitted in
any form or by any means, electronic or mechanical, including photocopy, recording,
or any information storage and retrieval system, without permission in writing from
the publishers.

British Library Cataloguing-in-Publication Data
A catalogue record for this book is available from the British Library.

ISBN 978 1 86126 952 2

Designed and edited by Focus Publishing
11a St Botolph's Road
Sevenoaks
Kent TN13 3AJ

Printed and bound in China by 1010 Printing International Ltd

CONTENTS

ACKNOWLEDGEMENTS

As with *Digital Camcorder: A User's Guide* (The Crowood Press, 2006), the writing of this book was a very isolated process for its author with almost all outside communication and assistance being provided in cyberspace via the internet and email. As a result, my circle of real flesh and blood contacts has dwindled, while my list of virtual (and imaginary) friends is now the stuff of legend. The streamlined, no-nonsense approach that I try to convey towards non-linear editing and the supporting technology is the result of years of brainstorming and technical problem solving for newsstand magazines – particularly in the bad old days when DV technology was in its infancy and simply getting hardware and software to work reliably became the moviemaker's primary focus. It still amazes me how any videos got edited then, but I would like to thank Bob Crabtree – then editor of *Computer Video* magazine – for inviting me into that world, and current publications such as *Digital Video* and *Practical Digital Video* for keeping me active in it.

The product images used in this book were provided by the following companies: Apple Computer, Avid Pinnacle Systems, Dell, Packard Bell, Grass Valley, Crucial, Intel, Nvidia, Creative Labs, Seagate, Hewlett-Packard, LaCie, Viewsonic, Logitech, LG Electronics, Sennheiser, Rode and LogicKeyboard.

On a personal level, heartfelt thanks go to my little boy Logan, for keeping my priorities in check and reminding me that the local playground is indeed a more exciting place than my pokey home office. As always, my wife Margaret seems to provide enough strength, sanity and self-confidence for both of us, and it is unlikely that I would manage to complete anything without her support.

1 INTRODUCTION

As I type this, my previous book, *Digital Camcorder*, has just been put to bed and is in the process of being printed and, as with that one, this book will probably be written in a linear fashion from front to back. In many ways it is a neat and structured way of working, but it fails to take advantage of the immense freedom offered up by today's word processors and is very different from the approach I normally take to video. As you will discover in the following pages, video editing on a home computer is granted many of the same freedoms that you would now come to expect from word processing. Sequences of video clips can be rearranged with a few mouse clicks in much the same way that you would reorder words or paragraphs in a word processor. Mistakes can (normally) be corrected at any stage of the editing process, and visual effects serve to emphasize key points, just as you might apply bold, italics, underline or coloured highlights to text.

Non-linear editing (which will be referred to as NLE throughout this book) is a relatively new development for video makers, and possibly the most important thing to happen since the creation of the camcorder itself. In the days before computer-based cutting, video was edited by copying clips in sequence from one video tape to another (in fact, the earliest methods of video editing in a broadcast environment involved cutting and splicing the tape itself, but that long predates video's entry into the home market). Editing suites were often complex Frankenstein beasts, composed of two or more video decks for playback and recording, a central controller, a vision mixer and an audio mixer. There were often computers too, taking control of the machinery and enabling an automated and precise assembly of predefined sequences. With all this high-tech hardware in play, edit suite hire was expensive and programme makers would go through a simple 'offline' process first – cutting their project in a simpler (and cheaper), two-machine environment with no effects, sound mixing nor any of the other careful tweaks and balances that smooth out and polish the finished piece. These creative cutting decisions would then be fed into the big 'online' edit suite, where the finished movie would be assembled. Because of the linear structure of video in those days, video had to be laid down to tape in sequence – from the first fade-in to the final shot, with all effects and titles applied as you went along.

The workflow behind video editing is quite different today, as scenes can be developed in any order, visual effects can be introduced after the main picture cut is complete and, if the final movie runs too long (or if certain scenes seem excessive or unnecessary), chunks can be removed with virtually no fuss whatsoever. Do not be fooled into thinking that this level of versatility is as new as computer-based cutting though – film editors have long enjoyed the benefits of non-linear editing, as they spend their time piecing together individual lengths of celluloid. In this case effects can be applied only later, once a show print is being prepared, and the restructuring or the removal of scenes is as simple as separating and reattaching the strips in their new sequence.

So, in some respects, computer-based NLE has brought us full circle, giving video editors much the same level of versatility in their working methods as film editors have. But even film editors have jumped aboard the bandwagon, since editing on a computer and sending their cutting decisions to the film labs for the preparation of a print is far cheaper and quicker than ordering physical prints of all shots and physically stringing them together. And, as we shall soon see, much of the methodology and terminology of physical film editing has now made its way into the world of video editing software.

NLE works by transferring video from the camcorder's tape, disc or hard drive to the editing computer – a process referred to as capturing (or sometimes 'digitizing'). These clips are arranged in sequence within the computer, just as you would paragraphs in a word processing document, and then sent back to tape or moved on to other publishing applications. The pioneer of computer-based, non-linear editing was Avid, who introduced the Mac-based Media Composer system into the high-end broadcast market in the late 1980s. I first got to work with one in the last of my four years studying film and TV production in Edinburgh – and, in those days, Avid was a name you used with great reverence. Everybody knew that this was the way forward for video editing, but the machines were so expensive and in so much demand that they

seemed to be accessible to only an elite and privileged few. As it would turn out, however, with so many people keen to hit the ground running with new and exciting technology, there were far more Avid editors in the freelance marketplace than there could possibly be work for. Avid's broadcast editing machines did (and still do) cost a substantial fortune, and, while any post-production house worth mentioning simply had to have Avid, the number of systems in use was still small compared with the number of people keen to work with them.

While broadcast-level editing systems remained far out of reach of small-time freelances and video enthusiasts, interesting things soon began to happen in the consumer arena, powered by the growing accessibility of home computers – be they Windows-based PCs, Amiga computers or Apple Macs. My first computer was bought in 1996 for a whopping £3,000 that I could barely afford. Most of that money was eaten up by newfangled devices such as scanners and a CD writer, and even at that time I made sure that it came with some form of video editing tools – which gave me the ability to bring low-quality footage on to the computer and edit it, but no means to send it back to tape. Storage space was limited too (the machine had a 2GB hard drive, which would barely house a computer's core operating system today) so to work on any big projects was far beyond my reach. Still, by the summer of that year I had created a nice interactive CD-ROM showreel to hand out at a production expo that ran concurrently with the Edinburgh Film Festival.

NLE's development in the consumer mainstream was fraught with peril – caused largely by the huge demands it made on relatively modest computers. My very limited, input-only VideoBlaster card was quickly superseded by a range of capture devices, providing reasonable quality video capture as well as the means to output back to a TV or video recorder. Two companies – Miro and FAST – were market leaders at the 'prosumer' level, with their DC30 and AVMaster boards. At the home-user level, great strides were being made by the graphics company Matrox with its Rainbow Runner hardware, which piggybacked video capture hardware on to the system's graphics card, delivering a very workable solution for just a couple of hundred pounds. Even then, the world was not rosy and bright. To introduce two or more random computer components can give rise to conflicts and unexpected bad behaviour, and this was very much a problem when NLE was new. Hardware and software developers could not test their products with every combination and permutation of computer gear on the market, and so many keen editors found themselves spending more time hunting bugs and solving problems than cutting video.

And then there were quality concerns: moving video and audio from place to place via analogue cables will result in a loss of quality. The degree of that loss depends on the type and the quality of the cable itself, as well as the recording medium used to make the copy. Copying video to the computer then back out to tape would often result in a noticeable quality degradation, not helped by the big necessary evil of video editing on a home computer – compression. In order to handle the immense size and data rates associated with video, video capture hardware had to squeeze the information into more streamlined files. This often meant indexing and discarding colour information; but the more video was compressed, the more information was thrown away and the worse it would look. Hardware such as the DC30 and Rainbow Runner offered varying degrees of compression and did reasonable jobs for their price, but higher quality settings required large, fast hard drives for storage and playback – and, with computer

Getting DV equipment to work seamlessly with computers and editing software was a great challenge before the technology became standardized. One of the most successful players in these turbulent times was Canopus with its legendary DVRaptor.

technology still maturing, these were horrendously expensive.

Things finally started to move in a sensible direction with the creation of DV and MiniDV camcorders. Aside from the small scale, high quality and relatively low price of the camcorders themselves, they delivered a huge advantage for the editor in their ability to feed footage in its native digital form via a digital FireWire cable. With DV, footage is compressed in the camcorder itself before being written to tape, and what comes through the FireWire cable is a direct copy of the tape's contents, with no need to unpack and crunch it again. What is more, the video is compressed more efficiently than most analogue capture boards could manage – a movie crunched to the same degree by the editing system would look nowhere near as crisp and detailed as the MiniDV equivalent.

But the DV revolution did not happen overnight. First, many existing camcorder

9

The Apple iMacDV is a perfect representative of mainstream computer systems coming ready for DV editing straight out of the box.

enthusiasts were already kitted out with analogue camcorders and reluctant to shell out thousands of pounds (as was then the case) on the new format. And the trials of getting footage into the computer continued, as developers attempted to provide working FireWire connections to computers with varying success. And as if that was not bad enough, silly import tax laws in the European Union saw to it that the FireWire ports on camcorders coming to these shores were capable only of sending video out but could not receive it. By contrast, camcorders in the USA, Canada and Japan could be used as player and recorder. Those in the United Kingdom and mainland Europe could act only as players, leaving any editors with working systems an added dilemma about how to archive their movies when they were finished. One solution was to buy an expensive digital video recorder, and another was to hack their camcorder's software with one of the many DV-in enabling devices that came on to the market over the following few years. These hacks were the saviour of many an enthusiast's video aspirations, but

they did invalidate the camcorder's warranty in the process (and were more recently deemed illegal and forced off the market altogether). Thankfully, consumer pressure has now ensured that the vast majority of MiniDV camcorders sold in the EU have working DV inputs, but some of the cheapest budget models are still 'nEUtered', so buyer beware!

On the editing side, the biggest watershed came when Apple and Microsoft decided to support a standardized 'OHCI' version of FireWire (and DV video) directly within their computer operating systems. Straightaway, mainstream computer hardware companies were able to sell inexpensive FireWire boards that actually worked, and software developers – who were once forced to bundle their wares with compatible capture hardware – were free to sell editing programs off the shelf, confident that they would work with the new breed of FireWire-equipped computer. Realizing the potential of digital video in the home and in business, Apple took things much further, equipping every new Mac with a FireWire socket, as well as pre-installed editing software called iMovie. And while iMovie started off as a poorly thought-out nuisance of a program, it has now matured into a highly capable home editor. Between it and the company's 'prosumer' offering, Final Cut Pro, Apple has effectively driven all third-party developers off the Mac platform, keeping it exclusively for itself. Its main competitor, Adobe, has only just recently started developing its 'Premiere' editor for the Mac again. Microsoft, on the other hand, has proceeded with a more open-door policy. Despite providing its own basic editor, Movie Maker, any Windows users taking an interest in video editing are encouraged to investigate the massive selection of programs that are currently on the market, from entry-level solutions costing under £50 to more advanced offerings ranging upwards from £300.

While hardware was the main focus of editing systems in the 1990s, the new century has seen software move to the forefront. Just about any home computer bought today is powerful enough to handle the burden of DV editing, and hard drives are now so spacious, so fast and so cheap that they require precious little consideration by the user other than to wonder how many he should buy. The new ease with which people were able to get video in and out of the computers spurred hardware manufacturers to push the envelope further. They addressed two issues: the remaining need for analogue AV connections and the fact that most systems required users to wait for special effects to process before they could be watched. The new breed of capture board would input and output analogue signals as well as DV via FireWire and display many effects immediately without rendering (most often on the computer screen and via analogue outputs only, while digital output still required rendering). Today, hardware companies are able to capitalize further on the latest digital video formats to hit the mainstream – high definition HDV – which require much greater effort on the part of the editing system than was previously necessary for DV. Even then, as home computers become faster and cheaper, the need for dedicated hardware to facilitate real-time playback of effects or high definition editing is steadily diminishing.

Throughout the course of the following pages I shall discuss the technology that currently exists in the mainstream for editing digital video in most of the popular consumer formats. But aside from the kit itself, the bulk of the book will focus on cutting techniques, outlining the language of movie editing, giving practical tips on how to approach different types of edit and showing how to use the staggering array of special effects on offer (and telling you when to leave them well alone!). There is a lot to say and, just like *Digital Camcorder* before it, I make no apologies for concentrating purely on this one aspect of the process.

High-definition video, as shot with the pictured Sony FX1 camcorder, is ushering in a new chapter in consumer and prosumer-level NLE. The format brings with it a good many hurdles, but the early teething pains are nowhere near as severe as they were for DV pioneers.

2 THE NON-LINEAR EDITING SUITE

WHY EDIT?

To continue my analogy of the word processor: editing video is the equivalent of adding punctuation and paragraph breaks to text. It is a process of organizing ideas and presenting the content in a form that is easy to read and, ideally, engaging too. Applying commas to a sentence helps to provide focus for its context and meaning, and dividing sentences with full stops helps to isolate and emphasize individual concepts and ideas. How long and verbose – or short and curt – a sentence is also affects the perceived tone of voice, appearing relaxed and methodical or breathless and urgent. All the same concerns are at play when working with video. It is more than likely that the raw footage you've recorded to tape is a visual equivalent of busy, unfocused babble. Editing helps to make sense of it all. Video has a language of its own with many widely accepted structural and grammatical qualities. With this in mind, the objective of your edit becomes clear – you need to trim fat, present ideas efficiently and keep the project moving at a pace that is appropriate for the subject and the tone you want to create.

Before editing tools became as affordable and accessible as they are today, many newcomers to video tried to edit their movies 'in camera' during the shoot itself. They would storyboard their projects meticulously, then shoot every shot in sequence from beginning to end. It is a brave thing to attempt – with very few exceptions I would even say lunatic – but fraught with problems, such as the camera operator's reaction times in hitting the record button, pressure on the on-screen talent to get everything right on the first take, as rewinding and reshooting shots further compromise the accuracy of your edit. And then there is the lack of versatility in being able to restructure the movie or tighten up scenes. For control, precision and preservation of sanity, your best method of approaching a movie is to shoot the best footage you can get and worry about the cutting and structure later in a separate editing process. Of course, if you've already read *Digital Camcorder* you will be aware that there are considerations to take into account at the shooting stage to make editing easier, but setting simple goals at each stage – such as shooting strong visuals, recording clean sound and cutting fluid video – will yield much better results than trying to do everything at once.

HOW NON-LINEAR EDITING (NLE) WORKS

Linear vs Non-linear Editing

Linear editing is a process in which video and sound clips are assembled in sequence from the very first to the very last, applying visual effects and titles as you go. In a traditional

Avid was the leader in developing non-linear editing technology for the broadcast industry. It is still the main driving force at the high end, but has also bought a large share of the consumer and prosumer market.

sense, it means assembling a sequence through tape-to-tape copying. All video was edited in this way before NLE became available, and, even then, only the most affluent production houses could afford non-linear edit suites. Today, however, the new NLE technology is so affordable that many video enthusiasts have editing systems in their homes that ten years ago would have cost hundreds of thousands of pounds. In non-linear editing all video, audio and still images are brought on to the computer's hard drive and arranged into an edited narrative using dedicated software. Unlike the old practice of linear editing, clips can be shortened, lengthened or rearranged at any time. Editors can work on individual sequences in isolation, then string them together into a larger movie, and special effects, such as transitions, video filters and colour correction, can be applied after the main picture cut is complete. Once finished, the edited movie is sent back to tape (often to the same camcorder that was originally used to feed video into the computer) or shared as a DVD, multimedia presentation or as streaming video on the internet.

Avid for Avid

When NLE first came into being the demands of the job were often far too tough for the computers of the time. They were weak and very expensive by today's standards, and

13

fundamental requirements such as spacious hard drives in which to store all your media pushed system prices up to astronomical levels. For some years video editors would use their splendid Avid editors for decision making and 'offline' editing with low-quality data files, but assemble the final high-quality cut in a traditional linear suite. Only the most well-heeled establishments could afford a fully-featured online Avid suite. But computer technology becomes faster, better and cheaper every year and so it was not long before NLE replaced the old linear suites completely. A simple home computer with an inexpensive entry-level editing program is now capable of doing more than many traditional linear editing suites with their racks upon racks of expensive mixers, consoles and controllers.

Video capture requires a means of connecting a camcorder or video player to the computer by using either a FireWire port built into the computer itself or via an add-on device such as this.

A FIVE-STEP PROCESS

The video editing process may be distilled into five basic stages:

Capture

Video capture (sometimes known as digitizing if you are starting off with analogue tapes) is the process of copying footage from tapes to a computer's hard drive. Today, most consumer video formats are captured via FireWire connections (more on that later in this chapter), but some editing systems also provide analogue sockets for connection to video recorders, DVD players and older camcorders. A new breed of tapeless camcorders now exists, however, which use hard drives or recordable DVD discs, and, in these cases, video can be transferred to computers via USB channels too. As media technology matures and people start using pictures, sound and video in different ways, there is also a growing market for video recording on portable devices such as mobile phones, and the quality of these recordings is

getting better all the time. While the video formats used by phones are very different from those of camcorders, editing software developers are now beginning to support them too, allowing video enthusiasts seamlessly to mix footage from camcorders, phones and even digital stills cameras. These multimedia files can be copied direct from phones and cameras via USB or stored on memory cards and slotted straight into the editing system if it has a compatible reader. Whatever the format, media and interface between camera and computer, the result is always the same – getting your footage on to the computer for editing.

Assembly

Once on the computer, video clips must be organized, trimmed of unwanted and unnecessary frames and pieced together into a coherent narrative sequence (or incoherent, abstract montage, if you're that way inclined). With few exceptions, the trimming and cutting is a non-destructive process – meaning that the video files themselves are not actually split, and the unwanted tops and tails trimmed from each clip are not really deleted from the system. In the bad old days this used to have the disadvantage of not freeing up any precious hard drive space, and only the most advanced editors provided the option of purging the system of discarded or unused media. Today, hard drives are cheap and immensely spacious and so this is much less of an issue. On the plus side, non-destructive editing is a huge bonus in allowing users to correct trimming mistakes at any time, knowing that nothing has been lost. It also allows them to work on several alternative cuts of a project without the need to duplicate files on the system.

Sound Mixing

Sound is possibly the most overlooked part of video production and postproduction. Well-recorded and well-mixed audio can make very average visuals seem polished and professional, but poor sound will make even the best video and the most innovative picture edit seem rushed and amateurish. As you move through the edit process, you will find that the sound and the picture acquired by your camcorder can – and often should – be treated independently. Cutting picture and sound at different times helps to maintain a sense of seamless fluidity, which will be further enhanced if you've been clever enough to record ambient background noise (called 'wild tracks') during the shoot. The appropriate use of music can accentuate the mood you are trying to convey, but inappropriate music can brutally flatten the whole project in one swift blow. Remember too, that if you are editing for playback on DVD some editing programs allow you to mix surround sound as well as stereo. On a technical level too, there is the need to make sure that the combined sounds being used in a mix are well balanced to provide a natural sound, and to make sure that the finished soundtrack isn't too loud – digital audio that is cranked too high will sound absolutely horrible.

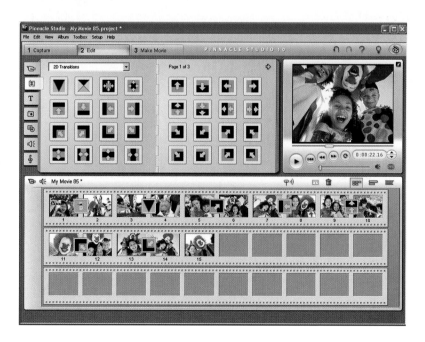

Video editing is typically a non-destructive process. Beginners' editing programs such as Pinnacle Studio, pictured here, make simple assemble editing easy with a storyboard interface as well as a timeline.

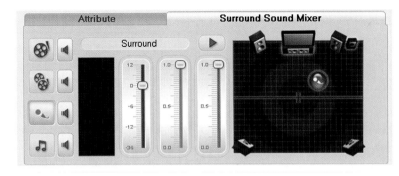

Even many of the most affordable editing programs – such as Ulead VideoStudio – offer well-featured sound editing tools, including surround-sound panning.

Effects

For the most part, video effects are by far the most overhyped part of the editing game and largely because they are the easiest features for hardware and software manufacturers to demonstrate and sell. If you ever visit a digital video trade show you will see all the NLE manufacturers and dealers demonstrating their wares with the use of crazy effects that make the image flip, flop, twist and bound around the screen. And the sad thing is that many of the newcomers that turn up to

Name	Type	Length	Comment 1
▶ 🗀 Favorites	Bin		
▼ 🔒 Video Transitions	Bin		
▼ 🔒 3D Simulation	Bin		
🎬 Cross Zoom	Video Transition	00:00:01:00	
🎬 Cube Spin	Video Transition	00:00:01:00	
🎬 Spin3D	Video Transition	00:00:01:00	
🎬 Spinback3D	Video Transition	00:00:01:00	
🎬 Swing	Video Transition	00:00:01:00	
🎬 Zoom	Video Transition	00:00:01:00	
▶ 🔒 Dissolve	Bin		
▼ 🔒 Iris	Bin		
🎬 **Cross Iris**	Video Transition	00:00:03:00	
🎬 Diamond Iris	Video Transition	00:00:01:00	
🎬 Oval Iris	Video Transition	00:00:01:00	
🎬 Point Iris	Video Transition	00:00:01:00	
🎬 Rectangle Iris	Video Transition	00:00:01:00	
🎬 Star Iris	Video Transition	00:00:01:00	
▶ 🔒 Map	Bin		
▼ 🔒 Page Peel	Bin		
🎬 Page Peel	Video Transition	00:00:01:00	
▶ 🔒 QuickTime	Bin		

Transition effects are bountiful in almost all video editing programs – from the most inexpensive to advanced offerings, such as Apple's Final Cut Pro, pictured here.

Video filters alter the overall appearance of footage. They can be used for subtle effects or for a more stylized result as seen here.

browse come away thinking that this is what editing is all about – and then shell out ludicrous amounts of money on real-time effects cards and advanced software that they may never find a use for in the real world. Going back to our text analogy, video effects are often the equivalent of the use of **bold**, *italics* or <u>underline</u> in a document. But, regrettably, while most writers use such effects to emphasize a specific point and to draw attention to a particular idea, many inexperienced editors use video effects purely because they can – drawing the viewer's attention only to the effect itself. You **would** become *very* annoyed to read a <u>book</u> that was *PEPPERED* with <u>**this**</u> **type of thing** from start to finish, but that's effectively what many do by adding wild transitions and video filters for the sake (or fun) of it. That is not to say that effects can't be useful – and, of course, there are video enthusiasts who rely on effects for more abstract and expressionist work – but for straightforward, narrative projects, they should be handled with caution and restraint.

At the basic level, there are two types of video effect: transitions and filters. Transitions sit between clips and create a gradual change from one to the other. The most commonly used one in film and TV is the dissolve, in which one clip fades away as the next fades in. There are also wipe effects, in which one clip is wiped away, revealing another beneath, an effect commonly used to introduce new scenes in the *Star Wars* movies. And there are more crazy page peels, spins and twists – in fact, just about any video editing program will dazzle you with its range of wild and wacky transition effects. Video filters are effects that change the appearance of a clip as a whole. There are subtle and very useful effects, such as colour correction and image grading tools that correct colour cast problems in the source footage and help to enhance a specific look for emotive effect, and there are compositing effects that allow several videos or image files to be combined into a single clip. But there are also such outrageous effects as watercolour filters and mosaic patterns which, for the most

17

part, belong in a toy box rather than a serious toolkit.

Audio effects deserve a mention too. As with video effects, there are transitions and filters. There is a limited range of transitions, but the one you will find yourself using almost exclusively is a simple audio dissolve, allowing one sound to fade out as another fades in. Audio filters are vast, however, and, unlike the video equivalent, many of them are genuinely useful, such as compressors, noise gates and equalizers – all of which help to boost or manipulate sound in subtle but persuasive ways. We shall be exploring these in greater detail in the coming chapters.

Export

Once a movie is finished you will need to get it out of the editing software (and possibly off the computer altogether) so that it can be easily screened. A few years ago the only serious options were to send it back to tape, and make VHS copies for distribution. Today, the options are vast, including DVD, multimedia presentations, export to personal video and audio players (or even mobile phones) and streaming on the internet. In this book we shall touch on the basic options, but in-depth guidance on the how, what and why of video publishing is stuff for another book altogether.

Analogue and Digital Video

With analogue video and audio, sound and picture are stored on tape as electromagnetic waves. Differences between wave frequencies and amplitude are smooth and gradual. Analogue formats are also prone to hiss, noise and other forms of mechanical interference, however, and the media on to which recordings are made can have a bearing in restricting the range of frequency and the amplitude of the electronic wave – effectively limiting the picture quality and the sound depth. Analogue video formats include VHS, SVHS, digital8 and Hi8 camcorder formats, together with traditional terrestrial broadcast signals. Digital video, on the other hand, breaks down images and sound into small data elements. Images are reduced to pixels and sound is encoded as a collection of samples. Digital video and audio are less prone to noise compared with analogue recordings, but with information broken down into neatly-defined blocks, the ultimate quality must depend on the number of samples and the amount of information afforded to each one. In the case of digital imagery, the key factors are the number of pixels that compose a picture and the size of the colour palette available to it. Sound quality is determined by the number of samples per second of audio (sampling rate) and the amount of digital information in each (sample size). So 'digital quality' does not really mean much, regardless of how many times the term is used in consumer electronics. There's a lot of low-bandwidth, streaming video on the internet that looks and sounds terrible, but it is still 'digital'. Fortunately, camcorder and home movie formats have improved immensely with the birth of the digital age. MiniDV and DVD, for example, are streets ahead of Video8, Hi8 and VHS. And, despite the ongoing argument over whether CD or vinyl is better (it's vinyl, and we all know it really...), but there's no denying that the CD is ultimately much more convenient and immensely preferable to the audiocassette.

DIGITAL VIDEO FORMATS AND COMPRESSION

Video can be stored and played on computers in many different formats. In fact, there are far too many video formats out there – behind every video format is a software development company earning licence fees from the supporting programs and products. Big companies such as Sony, Apple, Microsoft and Real Networks all have their own formats, and there are plenty more jockeying for position in multimedia and net streaming applications. Different video formats have different uses though, and most of the choices exist at the publishing stage. What defines each one, however, is its method of compression – and compression is one of the most important factors in the digital video arena.

In their raw state, video files stored on computers are big. Even a short movie will occupy a lot of hard drive space and place immense demands on the system for playback, as massive amounts of data are read every second. To make video more manageable, it is necessary to compress it into a more streamlined and lightweight form. But as it is impossible just to miniaturize digital bits and bytes, our only option is to throw information away. Obviously, there's some clever programming at work, with codes assigned to the video file in an attempt to index the lost information so that it can be recreated on playback. For every Encoder (which crunches video into lightweight streams) there must also a Decoder, which reads the file, unpacks it and interprets it for display – hence the word CODEC, which is derived from enCODer and DECoder. Video compression generally falls into two categories – interframe and intraframe compression.

Intraframe compression preserves each video frame as a complete, unique, self-contained image. In the case of DV, the amount of data required is reduced by reducing colour information rather than luminance (or brightness) detail. The approach lends itself well to editing since it allows cuts to be placed on any frame without the need to reprocess footage. The disadvantage is that video can be compressed only so far before compromises in quality become noticeable. In the days before DV, analogue capture devices often used Motion JPEG (M-JPEG) compression. As with the JPEG images used on the internet, file sizes could be made very small, but sharpness was compromised and detail became muddy in the process. To preserve quality but use less data, a different method of compression needed to be found.

Interframe compression works quite differently, by comparing adjacent frames within a movie and updating only the things that have changed from one to the next. In many cases a video will be composed of self-contained index frames (or 'I-frames'), placed at intervals of roughly 1 to 1.5 sec. Between these index frames will be a collection of Predictive ('P') and Bidirectional ('B') frames that contribute to the job of updating portions of the moving image. This IPB frame structure, often referred to as a Group of Pictures (or GOP for short), can result in superb video quality at very low data rates, but the fact that it is composed of partial frames rather than complete images can make it more difficult to edit. Until recently, the accepted rule was to edit intraframe formats such as DV and publish the finished movie as an interframe format such as MPEG-2, used for DVD. As high-definition video enters the mainstream, there is a greater need for high-quality, low-bandwidth formats at the editing stage. Thankfully, the editing of software is becoming more sophisticated and computers more powerful, so rules are now a little more relaxed concerning what you can and cannot edit.

19

RESOLUTION: STANDARD DEFINITION, HIGH DEFINITION AND MULTIMEDIA

Digital images are composed of pixels, which can be thought of as tiles in a mosaic. Each pixel has a regular shape – square or rectangular – and a single colour or tone. Seen from a distance, these pixels form part of a larger image, and the more pixels contribute to that image, the more detail it will have. A video's resolution is normally measured in terms of the number of pixels that make up its width and height. In theory, video can be created at any resolution, but some specific standards apply in the real world for different methods of delivery. TV sets are designed to support the standard resolutions adopted by broadcasters. The size of analogue broadcast images has remained unchanged for decades and is referred to as Standard Definition. Standard Definition television sets have a resolution of 625 horizontal scan lines for the PAL video standard (NTSC sets in the USA, Canada and Japan have 525 scan lines). Scan lines are the lines created in a cathode ray tube in traditional television sets. Modern LCD and plasma screens display pixels rather than create scan lines, but they are still intended to support specific broadcast resolutions. The resolutions of older analogue video formats are often referred to in horizontal lines, and of the home video formats that existed before digital media took hold, few came close to matching the image potential of the sets they were viewed on. VHS video delivers a resolution of around 250 lines, while SuperVHS and Hi8 pushed things a little further to over 400, so long as good connections were used to connect the player to a TV set.

High-definition video has become an exciting buzzword in the home cinema market in recent years, but the concept was first introduced back in the 1970s, with the proposal of a TV system called MUSE. MUSE, and other high-definition recording and broadcast standards were developed throughout the 1980s and the 1990s, and as well as being intended as a step up in broadcasting quality, they were also pitched at movie makers as a cost-effective alternative to film. Broadcasters were unwilling to upgrade, however, as a higher quality image would require more of their available bandwidth. This was a time in which satellite and digital TV were becoming a reality, and broadcasting companies were more interested in squeezing lots of poor-quality channels into their allotted bandwidth than delivering only a few channels of pristine, high-definition programming. More channels means more air time for advertising, and that has always been a bigger concern than programme quality.

High-definition video has finally made it into the consumer mainstream, thanks to the new low prices of LCD and plasma screen TV sets which can support much greater picture resolutions than standard broadcast. Naturally, the companies behind these TV sets are also keen for us to buy new DVD players, games consoles and camcorders in order to capitalize on the potential of our new TV sets, and the broadcast industry is finally following with high definition TV. The move to HDTV is not as inconvenient as it would have been ten years ago, however, compression techniques for digital broadcast have come along in leaps and bounds, allowing us to compress high-definition video into relatively tiny data streams. And then there is the fact that existing digital TV viewers are realizing that most of the channels they are currently being offered are utter rubbish – something new needs to be introduced to keep them watching!

As we have come to expect in the video market, there isn't just one high-definition video standard to choose from, broadcasters and high-end movie makers have expensive,

Comparative frame sizes. Top to bottom: standard definition video, 720p HDV and 1080i HDV images (DV and 1080i HDV frames have been reformatted to meet their intended aspect ratios).

data-rich formats such as HDCAM to work with, while we in the budget arena have HDV, which comes in two flavours – 720p and 1080i. The 'i' in 1080i stands for 'interlaced', meaning that image resolution is split into two parts, using alternate horizontal lines. So instead of 25 images per second at 1440 × 1080 pixels, you see 50 at a resolution of 1440 × 540. In 720p, the 'p' stands for 'progressive'. Images aren't interlaced and each frame is displayed in its entirety at a resolution of 1280 × 720 pixels. Which format is 'best' is still up for debate. The interlaced nature of 1080i is more motion-friendly than a progressive format with a lower frame rate, making movement and action

21

appear much more fluid. The higher number of image samples is also useful when converting between PAL and NTSC video standards. On the other hand, many people prefer the look of progressive video and consider it more 'cinematic'. Others contest that progressive footage is best for conversion to a film print. It should also be mentioned, though, that some new 1080i camcorders are capable of recording a single frame to both sets of interlaced fields, giving similar results to progressive scan and serving just as well for transfer to film. The two HDV standards reflect competing broadcast specifications which are being rolled out in different parts of the world, but we also cannot ignore the lucrative licensing franchises that are up for grabs for whoever can develop the single de facto consumer HD video standard.

Regardless of which flavour you choose, HDV is recorded to standard MiniDV cassettes and the video signal has the same data rate as ordinary DV – despite the fact that 1080i HDV has four times the picture area of standard definition DV. This is achieved by compressing HDV using interframe MPEG-2 compression. The resulting video looks great, but requires a lot more power from editing computers and sophistication in the editing software itself in order to cut it accurately and quickly.

Outwith standard definition video, high definition, DV and HDV, there is an almost limitless choice of formats and standards when publishing for multimedia projects or the internet. In many cases video is made to a much smaller frame size – often a quarter of the standard definition resolution or less – and frame rate can also be reduced to fifteen or twelve frames per second. Choice of codec also depends on the way it is being delivered. For CD-ROMs, for example, MPEG-1 or highly compressed QuickTime is often the most appropriate option. For delivery as internet downloads many now use DivX or MPEG-4

QuickTime files, while for streaming video (which is never actually downloaded to the viewer's computer), RealVideo, Windows Media and Macromedia Flash video are becoming the established standards.

AVI and QuickTime

AVI (Audio Video Interleave) and QuickTime video formats are the standard video file types used by Windows and Macintosh computers. Both formats offer immense freedom in frame size, frame rate and even the choice of codec used to compress them. And while they are closely tied to particular computer platforms, things have developed well enough so that they are often interchangeable and readable on either type of computer. The video types are recognized by their file extensions: '.avi' for AVI and '.mov' for QuickTime. When it comes to mainstream editing formats, such as DV, AVI and QuickTime, files are essentially identical, differing only in the way audio is interleaved with picture and the files themselves are presented to the player software. We see greater differences between the two formats when footage is compressed to greater degrees for multimedia or the internet, as certain codecs are supported in different ways.

MPEG Standards

MPEG stands for Motion Picture Experts Group, and the first incarnation of the MPEG video format, MPEG-1, was developed to allow movies to be stored and transferred easily within a very small bandwidth. The picture was often scaled down to a quarter of the area of standard definition and, as with all subsequent developments of the MPEG standard, it uses interframe compression, updating only the parts of the image that have changed between index frames. Limited as it is compared with full quality broadcast, MPEG-1 became hugely successful as the compression format for VideoCDs (VCDs) which, until the dawn of DVD, proved more popular than VHS tape for movie distribution in the Far East. At its best, the quality of VCD video is comparable to that of VHS.

MPEG-2 is the standard used for DVD Video discs and represents a huge stride forward from the limitations of MPEG-1. Video is typically created at full standard definition resolutions and is now being pushed to high definition through HDV formats. As well as providing compression for DVD, MPEG-2 is also used in digital broadcasting and in camcorder formats such as MiniDVD, HDV and hard in drive-based camcorders.

The third step in MPEG's development is MPEG-4 (MPEG-3 was intended as a high-definition video standard but never really got off the ground). MPEG-4 provides compression for all aspects of video publishing, from highly-compressed multimedia and internet delivery to high-quality, high-definition archiving and broadcast. It was quickly adopted by Apple as its codec of choice for downloadable QuickTime files and was also cracked and adapted to create the now infamous DivX video format, which has done for online movie sharing what MP3 did for music. MPEG-4 is continually being developed, however, and currently exists in basic and advanced profiles – the newest and more advanced option being known as H.264, which is being promoted as the codec of choice for creating high-definition DVDs and has also been picked up by Apple as its new compression tool for everything and anything.

AN ANATOMICAL GUIDE TO THE EDITING COMPUTER

At its best, an editing computer should have no more influence over your work than your desk or your office chair. You will be working close-up with some very exciting software, but the computer on which it is running should hardly even be noticed. In general, we become aware of the hardware in front of us only when it misbehaves, crashes, slows to a crawl, reboots itself without reason or freezes. Regardless of anything you might have been told, however, this is not acceptable behaviour for a computer of any type and should be avoidable if you take care.

Down to Basics – What Is a Computer?

Many of the problems newcomers face when they start editing video do not come from their editing software, their video hardware or the actual editing process itself, rather they are spawned from a lack of familiarity with computers in general. If you are already familiar with the workings of computers and Windows or Mac operating systems, feel free to skip this next little bit. If, however, your introduction to DV editing also coincides with your first experience of computers, read on.

In a nutshell, computers are big, powerful calculators that do lots and lots of sums every second. Every task you might ask a computer to perform is broken down into a maths problem and, for that reason, computers can only work with actual values and definitive answers. They do not recognize ambiguities and therefore are not capable of actual thought. They can do only what they're told to do, whether that instruction comes from you as the user, or the programmer who wrote cause-and-effect scenarios into the software you are using. Computers are composed of many parts, most of which will be introduced and explained in the following pages, but what people generally recognize are the main computer body (often an ugly beige box), a monitor (the screen), a keyboard, and a mouse. Some computers, such as Apple's iMacs, deliver the monitor and computer body in a single unit, and look quite stylish too. Mac and Windows computers are perfectly capable of editing video, although enthusiasts in both

ABOVE: A desktop computer made by Dell...

RIGHT: ...and one made by Apple.

LEFT: Even the majority of modern laptops are capable of handling the workload of DV editing.

camps will tell you that their side is the better. In order to tell the computer what to do, and understand what it's doing, an operating system needs to be present; the two best-known systems are Windows and Mac OS. Both options provide the user with a visual working environment in which utility programs, such as word processors and video editors, can be run. They also help to organize information on the system's hard drives (more about them later). Every piece of information – be it a video clip, a photo or a text document – is saved to the hard drive as a 'file'. To ensure that things don't get too cluttered and chaotic, users can create organized sections within the drive known as 'folders' or 'directories' – the terminology and visual representation are not dissimilar to the typical organization of a filing cabinet or lever arch file, except that computers automatically order things alphabetically, so you need not bother. To the general user, a particular directory might be thought of visually as a little folder in the main hard drive, for example. But the computer will see things in a different way, often referring to files and their locations in a single text line, such as

C:/videos/holidays/lille01.avi. On Windows computers, drives are assigned letters to identify them, so that little piece of script tells us that the video clip called lille01.avi can be found in a folder called 'holidays', which resides inside another folder called 'videos', and that in turn is located on the main hard drive, which is given the letter 'C'. Incidentally, the computer's main hard drive (the one that contains the operating system and software), is almost always given the letter 'C'. 'A' and 'B' are traditionally reserved for external media devices such as floppy disc drives.

For most jobs a keyboard is reserved for inputting text and numbers, while the mouse is used as a pointing device to select elements on screen or browse visual menus within a program. As we shall soon see, however, the keyboard can also be used for controlling most of the features within a video editing program, and to work in this way can make the job much quicker. Keyboards and mice are also used in different ways for playing games, but that's a

different story altogether, and not a practice I endorse since computer games steal all your time away and leave you with nothing to show for it but repetitive stress injury. Be warned!

The remainder of this chapter will cover the basics of buying a computer for video editing (or building one yourself, if you're feeling brave), as well as explaining the whats and whys of each of its main component parts. Once you realize that computers are essentially just a collection of self-contained bits and start to understand what each of those bits does they won't appear quite so intimidating.

What to Spend – What Do You Plan to Do?

All too often I can be found haunting video trade shows, often representing related magazines or publishers. There I meet a lot of video practitioners and a proportionally larger number of potential newcomers. Of the novices, all seem to have two things in common: first, they are all desperate to spend as much money as they possibly can and,

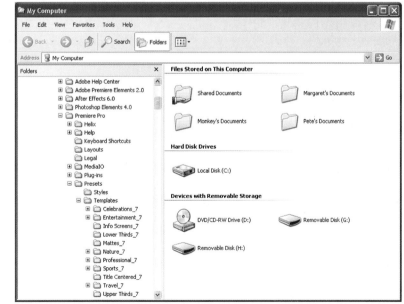

A typical hard drive folder structure as seen on a Windows system.

secondly, they think that owning expensive equipment will make them 'professional' video makers, as if the quality of their work will depend solely on the tools and not the craftsman. Of course, they don't like being told that they're on a fast track to insolvency and disillusionment.

It is true that the top-spec camcorders and expensive real-time editing systems are very seductive, but you really should stop and consider whether you actually need them before parting with any money. If you want to make money from video, think of your expenditure as a reward-based investment system. The more money you make from your work, the more you can afford to buy. So start small and buy only what you need rather than what you want. Your first objective should be to make your purchases pay for themselves and, as you earn more, allocate a small percentage to an equipment fund, allowing you to upgrade over time as your business grows. By then, you will also have a better idea of what kind of work you're doing and how you like to do it, giving you the means to make much more sensible and informed purchasing decisions than you would ever make as a novice.

In many cases, it can make good sense to shop around for good quality used equipment. You will be surprised just how much is available in the camera and camcorder line, and that should tell you something about this game. Behind many second-hand camcorders or video editing cards on eBay is an owner who isn't getting the level usage or paying work required to justify his own initial expense. A further word of warning though: used gear seldom comes with a guarantee so never buy it 'blind'. If you cannot see the product before parting with your money be sure that the seller has a good reputation and an acceptable returns policy. If warranty and peace of mind are of crucial importance, buy new.

Mac or Windows?

The computer world is, on the whole, divided into two camps – Windows and Mac users. Some computer enthusiasts choose to thumb their noses at both, in favour of systems such as Linux, but they are in a small minority. In fact, Mac users are also a minority when it comes to day-to-day home and office computing, with the lion's share of the computer market running off the Windows operating system. Many enthusiasts will insist, however, that Macs are the way to go for media work such as graphic design, audio mixing and video editing. Well, Mac enthusiasts will insist that this is the case, anyway. The truth is that just about any new computer bought today is up to the job of video editing regardless of whether the installed operating system is Mac or Windows. But that does not necessarily mean that you can choose randomly. Here are some key points to consider:

About Macs

Despite how things may seem at first, Apple does not actually make the component parts of a Mac computer. Just like a Windows PC, Macs are composed of components that come from a range of manufacturers. What Apple does, however, is to decide on a limited range of devices that can be used in their systems. Controlling the range of hardware that can go into a Mac helps to ensure that there will not be any conflicts between parts and that the operating system can be tailored to run efficiently with specific devices. It helps customer support too. Call a support centre and you will need to tell the operatives only two or three basic facts about an iMac before they can call up a full inventory of what's inside, because all Macs from a specific model and range are made the same. There is a much more controlled integration between hardware and software on the Mac platform

A high-end Apple Mac computer delivers immense speed and offers greater flexibility than more affordable Macs when it comes to hardware upgrades, but it is also one of the most expensive options you can consider.

than you will find in Windows PCs, and Mac enthusiasts will insist that this makes Macs more reliable and better performers. This isn't necessarily the case in the real world, but it does make the Mac route a much simpler one if you are in any way technophobic.

It is also worthwhile considering that Apple has a very active interest in the world of video editing. All new Macs come with FireWire ports with which to connect to DV and HDV camcorders, and a very useable video editing program called iMovie. Advanced video editors are well catered for with Apple's Final Cut Pro software and a lighter version – Final Cut Express – is provided for ambitious amateurs trying to bridge the gap between beginner and professional editing techniques. And while the programs are good, they are the only established ones available for the Mac, unless you are willing to pay top-dollars for high end offerings from the likes of Avid. Before Apple took a direct interest in video editing Mac users had a choice of programs from Adobe, Digital Origin and others. All those developers were driven out of the Mac market, so if you don't like the software Apple has to offer (and comfort is a big part of choosing an editing program) then you had no alternative. At the time of writing, Adobe is just

Video enthusiasts are well catered for at all market levels and all mainstream video editing applications for the Mac are developed and released by Apple itself.

27

relaunching its Premiere software for the Mac. Do not assume either that Apple's editing programs are bug-free and 100 per cent reliable, a quick visit to technical support messageboards on Apple's own website will show how stable (or otherwise) the current versions of iMovie or Final Cut are. To the company's credit, however, Apple's interest in the video editing market is such that users do not have to wait long for updates and bug fixes.

Aside from the comparatively limited choice of software available to Mac users, one of the factors that tends to dissuade potential buyers is price. Even the most consumer-orientated iMac can cost more than a Windows PC and these affordable, one-piece Macs are limited in their upgradability; it is virtually impossible, for example, to add internal hard drives, change the DVD writer for a newer model or upgrade the graphics card. If you think that you will want to keep your Mac serviced in this way, the only option is to go for a more expensive PowerMac.

Apple's trump card in the Mac/PC debate came in 2006 when Intel processors began to be used in Mac computers. With Mac and Windows computers using the same kind of processor chip, Apple quickly presented its users with the means to treat Intel G5 systems as dual-platform machines, enabling them to install Mac and Windows operating systems and switch between them, depending on the job in hand. This capability does not overcome issues of cost (especially as users would have to buy a copy of Windows off the shelf at its full price rather than cheaply with a new PC), but, for serious users, genuinely torn between the two platforms, it is an absolute blessing.

The Windows PC

Computers that run Windows operating systems are normally referred to as PCs, although the term 'PC' really stands for Personal Computer and can also include Macs or machines running less popular operating systems such as Linux. There is a good choice of operating systems besides Windows that will run on a normal PC computer, but most are used for IT applications such as networking. The mainstream developers of video editing hardware and software write exclusively for Windows, and, unlike the Mac market, there is a lot of fierce competition in the Windows-based arena, covering all levels from beginner to pro.

Windows PCs are manufactured by a large range of companies, such as Packard Bell, who made the pictured desktop system. This extensive competition helps to keep prices keen, but also introduces many more factors regarding system configurations and the range of hardware elements in use.

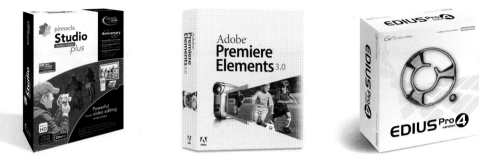

Unlike the Mac market, Windows users are faced with a diverse choice of software applications and development companies. Choosing can be a tricky task, but the range is often useful in a market where there is never a true one-size-fits-all solution.

Windows PCs have – rather unfairly – gained a reputation for being unstable and unreliable. It is true that the development of video editing on PCs was fraught with teething troubles, but you will often find that most of those who keep this reputation alive are dedicated Mac users who spend little time with modern Windows systems. I shall cover stability issues in more depth towards the end of this chapter, but a good Windows PC can be every bit as solid and reliable as a Mac, if you show a little restraint with the things you install on it.

There are several things that Windows systems have in their favour when it comes to video editing: First, the computers themselves can be bought or built relatively cheaply, and, as there are few one-piece solutions to rival Apple's iMac, even the most inexpensive PCs can often be upgraded to a good degree with additional hard drives, new disc burners, updated graphics cards and the like. And then there is the issue of choice: compared with the Mac market, the range of hardware and software products for Windows-based video makers is staggering, ranging from capture hardware for getting video on to the computer,

through accelerators to speed up performance, to software for editing, special effects and conversion of file formats. The choice can be daunting for the absolute novice, but those with experience of video editing will appreciate the opportunity to choose between products that best suit their way of working – in fact, in my case, I often find the need to flit between two or possibly even three programs when working on complex projects.

Windows PCs do not always come out of the box with video editing capabilities. Microsoft has devised its own free editing program MovieMaker as part of the Windows operating system, but it is a very basic affair, designed to whet the user's appetite for more advanced, third-party software. Also, the FireWire sockets necessary to get video on and off the computer are not present on all PCs. Many systems have FireWire sockets built-in, allowing them to send and receive digital video from DV and HDV camcorders, but those that don't will need an additional expansion card to provide them. Fortunately, upgrading a Windows PC for DV editing is a remarkably simple task if you can overcome the initial fear of opening up the computer itself.

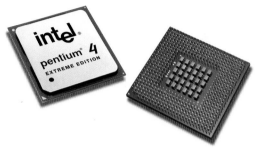

Computer motherboards act as a spine, providing a connection between all parts of the computer. The type of motherboard you choose will determine the kinds of component that can be connected to it.

Processors act as the computer's central brain and are responsible for most of the calculations done by the machine. Intel's processors (pictured) are currently the market leaders.

The Basic Core

At the heart of any Mac or Windows PC is a common array of components, each serving a specific task in the processing of data and the interpretation of commands. No matter how much you spend, nor which operating system and software applications you choose to work with, all mainstream computers do their job in much the same way with the same types of component.

Motherboard

A computer's motherboard (or mainboard) binds all its internal parts together. It can almost be thought of as the system's spine, serving as a central base into which all the other components are plugged. The motherboard provides connections inside the computer for a processor (or two), memory chips, hard drives, a graphics card, a sound card and a limited number of additional expansion cards, such as real-time accelerators for enabling the immediate playback of special effects. Externally, the motherboard also provides sockets for hooking

up a mouse, keyboard and other external drives and devices via FireWire or USB sockets. Windows PCs may also have serial and parallel ports for use with older equipment, but these connections have long since been lost from Apple Macs. In the case of some budget systems (and entry-level Macs, such as the iMac), graphics and sound chips are built into the motherboard rather than being provided as separate cards.

Processor

A computer's processor is often referred to as a CPU (Central Processing Unit). This is the part of the computer that runs calculations and carries out tasks dictated by the user and whichever program it might be running. If a computer were capable of thought, the CPU would be its brain. (But then, if computers had brains, they wouldn't need users or programmers.) In basic terms, the CPU works like a big calculator, running numerous and bountiful number of calculations at once. Much fuss is made about the speed of processors and until recently computer

30

enthusiasts would look primarily at the clock speed of processors, measured in hertz (Hz), or, more specifically, in gigahertz (GHz). In practice, however, the design and architecture also have a huge bearing on performance, and the game of choosing a CPU has become a rather confusing one for the novice. At the time of writing, there are two main processor manufacturers in the mainstream, Windows-based computing market: Intel (who control the largest market share) and AMD, whose processors are a popular choice with many computer fanatics. In magazines and internet forums you will see lots of graphs and benchmarks proving that processor X is faster than processor Y, and, on a purely academic basis of abstract number crunching, you could be convinced that the fastest processor is the best. In practice, however, it is not unusual to find that software developers have created and tested their programs by using only one type of processor (usually Intel chips for Windows-based applications) and conflicts arise when they enter the real word and are installed on different types of system. Macs use dedicated processors, made by whichever company Apple decides to commission. At the time of writing, Intel has just been given the contract to make CPUs for Macs and I expect this relationship to last for a good many years.

RAM provides the computer with memory, allowing it to unpack large amounts of data for processing. The more RAM a computer has, the more information can be made available to it at any one time.

RAM

RAM stands for Random Access Memory. In effect, RAM provides workspace into which data can be unpacked and edited before being saved on the hard drive. A lot of RAM allows a lot of data to be manipulated at once, contributing greatly to system speed and performance; 256Mbytes of RAM should be considered an absolute minimum for video editing, but 512MB or even 1GB and above are ideal. As with processors, RAM speed is also an important factor, determining how quickly data is accessed, moved and cleared. As with most things in the computer world, faster is better, but be aware that the system's motherboard will support only RAM of a certain type and speed.

Graphics Card

At its most basic level, a graphics card feeds a computer monitor with its visuals, providing users with a visual working environment. But, as well as connecting the monitor to a computer, graphics cards have an active role to play in video editing, as more and more editing programs offload the job of processing special effects to the system's graphics chips. Evolution has been faster in the graphics cards market than in any other aspect of computer hardware, pushed by the gaming industry and the demands of die-hard game fanatics. All these game-related developments also apply nicely to video though, making it quicker and easier for the computer to process colour corrections, video filters and transition effects. Some budget computers have graphics built into the motherboard, which is perfectly adequate for internet surfing and office work but potentially limiting for video editing. More powerful graphics processing is to be had from expansion cards. Graphics cards (and the built-in graphics chips on motherboards) have their own dedicated memory for image processing. Until recently, graphics cards used an interface called Accelerated Graphics Port (AGP) to connect to the system's motherboard. AGP gives faster and more direct communication between the graphics card and processor than was possible through normal PCI connections used for most other expansion cards. In recent years, however, AGP has been superseded by an even faster interface, PCI Express (PCIe). The very latest and fastest graphics cards are always prohibitively expensive, but, as graphics technology evolves so quickly, it is possible to pick up extremely capable cards at keen prices, if you are happy to go for last month's model instead.

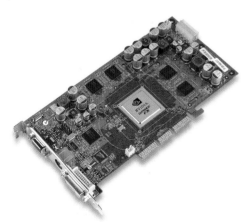

Graphics cards feed visual information to the computer monitor but also take much of the work away from the central computer processor when it comes to image and video editing. Many of the transition effects offered by video editing programs depend on the hardware acceleration provided by modern graphics cards, such as this one from nVidia.

Sound Card

Just as graphics cards and on-board graphics chipsets take over the processing of images, so sound cards handle audio generation and feed the computer's speakers. Unlike most graphics cards, however, sound cards enable input as well as output, allowing sound to be recorded to the computer via microphone and stereo analogue line inputs. As before, sound capabilities can be built into the motherboard or added with a dedicated expansion card. In this case, however, it is often fine to go with the system's own built-in audio (if available) for video editing, and many motherboards even offer built-in support for surround-sound. But there are advantages to spending a little extra on a superior card. Keen audiophiles will appreciate the fact that some cards allow recording and playback of very high quality, 24-bit sound, as well as digital ports for connection to devices such as MiniDisc recorders or even professional DAT decks. Professional sound cards also support balanced XLR sound connections and high-end digital interfaces.

Hard Drives

Hard drives provide storage space on your computer system. When non-linear video editing was an emerging technology, computer hard drives often had a low storage capacity and offered very slow data transfer speeds. To manage video effectively on a computer, editors had to invest in prohibitively expensive SCSI hard drives. Thankfully, things have greatly improved in the mainstream. Inexpensive ATA – and now Serial ATA (SATA) – provide incredibly quick data transfer and often have capacities exceeding 300GB, accommodating over 23 hours of DV or HDV footage. For an editing system it is often best to have at least two hard drives installed, one for the operating system and software and the second used purely for media storage. Working

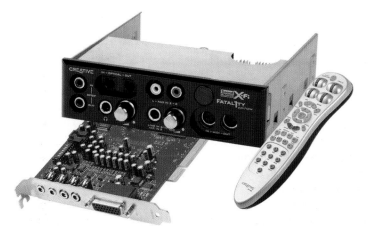

Sound cards feed audio to speakers and headphones, as well as providing a means of recording analogue audio on to the computer system. They can be very simple and inexpensive or, like this one from Creative Labs, more elaborate and feature-rich.

in this manner lightens the demands on your system's main hard drive and improves performance overall. Some editors add even more hard drives to their system for use only with rendered files processed by their software. It is worth remembering too that data stored on hard drives can become fragmented,

meaning that the computer might break files into several chunks of data to make the best use of its available space. This is a common occurrence if you are in the habit of saving and deleting small files, as happens automatically with common tasks such as internet browsing. System performance will be

Hard drives are used for the storage of files. Video files are big, and the more hard drive space you have available, the better. Ideally, you should have a second, dedicated drive for media files.

External hard drives, connected to the system via FireWire or USB 2.0 connections, are a good option if you lack the confidence to install an internal hard drive yourself.

compromised slightly if video files are fragmented over different parts of a hard drive, and so keeping a dedicated drive purely for media helps to ensure that files are self-contained and quickly accessed. Of course, it doesn't hurt to defragment the system and media drives on a regular basis. Windows systems have the tools for this built in to the operating system, while Mac users may need to buy dedicated software such as Norton Utilities.

Disc Drives
All computers now are provided with some kind of optical disc drive. At the most basic level, this will be a CD-ROM drive, but the popularity of DVD movies and DVD-ROM-based games has made DVD-ROM drives the new basic standard. You will find that many video editing programs are now supplied on

DVD-ROM too, allowing developers to include a rich selection of stock graphics and style templates, which you may or may not have use for. A video editing system, however, should really be equipped with a DVD burner, allowing you to make DVD video discs of the projects you edit. Choosing a burner can be a little confusing as there are two different recordable DVD standards: DVD-R/DVD-RW and DVD+R/DVD+RW. There are some issues of disc compatibility with older, set-top DVD players, but most modern players should be able to recognize and play discs of any type. To avoid confusion, though, I would recommend that you go for a multi-format drive which records to all DVD types. They are no more expensive than a single-standard burner and will allow you to choose your blank media freely. DVD burners can also record to CD-R and CD-RW. Some will also support double

DVD writers are becoming the standard type of optical disc drive for new computers. They are also inexpensive to buy separately and easy to install.

As with hard drives, external DVD writers are available if you are squeamish about the idea of playing with your computer's innards.

layer DVD+R DL discs, but, while their 8.5GB capacity is tempting for wedding videographers, who deal with long movies, I do not recommend the use of DL discs for video applications since the positioning of layer breaks is seldom managed well and many set-top DVD players have problems making the switch from one disc layer to the other.

Card Readers

Solid state media – that is, recordable media with no tape or moving parts, have become popular in the world of portable music, media-rich mobile phones and digital cameras. The cards themselves come in many formats, such as SmartMedia, Compact Flash, Memory Stick, SD Card and MultiMediaCard, depending on which company makes the device itself. Sony developed Memory Stick, for example, and Panasonic is behind SD Card, so both companies have a strong vested interested in pushing their own technology. It is very likely that you will want to use stills, audio or even video from your camera or phone in video projects, so a card reader is a good idea to provide easy access to the card. To avoid

Memory card readers often support several different types of card, freeing your choices when you buy a camera, camcorder, mobile phone or MP3 player, and also removing the need to connect each individual device to your computer in order to access files.

complication, most card readers support many different card types. Some are connected to the system via USB cables, while others can be installed into the computer case itself. Many laptops come with memory card slots built in.

Connections

All computers have an array of sockets for connecting them to external devices, such as printers, scanners, external hard drives and camcorders. Here is a rundown of the most common computer interfaces you are likely to find:

Monitor port Monitors are connected to the back of a computer, attaching to its graphics card or direct to the motherboard's sockets, depending on how the system is configured. Inexpensive monitors are normally fed analogue video signals via a VGA port, but more advanced graphics cards and flat panel displays use a Digital Visual Interface (DVI) for a clearer image and more accurate representation of digital graphics and video. If your graphics card is DVI but your monitor only has a VGA plug, small adapters are

---

Floppy Disk Drives

Floppies have been a staple part of Windows systems for many years but have recently started to become scarcer, largely because floppy disks have a tiny capacity of 1.44MB, which is fine for text documents but no great use for today's media-rich computer habits. A floppy drive is still useful for hands-on tinkering in the computer's guts, but most day-to-day users will have no real need for one. Floppy drives were dropped altogether from the spec of Apple Macs many years ago.

A VGA port is used to feed an analogue signal to older or budget displays.

DVI ports provide higher quality digital information to newer monitors.

available to convert the digital signal to an analogue stream cheaply.

USB USB stands for Universal Serial Bus and has become the most common interface for connecting scanners, printers, webcams and digital stills cameras to computers. Most of the small peripherals available for today's computers are connected via USB, and that can also include mice and keyboards. All modern computers have USB ports, regardless of whether they are Macs or Windows PCs, but there are now two versions of USB: version 1 enabled a relatively low data transfer rate of 1.5MB per second, making it fine for the transport of small files, or even large amounts of data if time is not an issue, but was

thoroughly unsuitable for real-time DV playback; USB 2.0, on the other hand, allows data transfer speeds of up to 60MB per second, making it much more suitable for external hard drives and video devices. USB 2.0 also supports older USB 1 devices and is now the standard on all new computers. Older systems might offer only the slower USB ports, either way, both versions can also provide power to connected devices.

IEEE 1394 IEEE 1394 is a generic term for the data transfer port on almost all DV and HDV camcorders. Apple calls the connection FireWire, while Sony calls it iLink. FireWire is the most commonly used term, however, and that is how I shall refer to it here. FireWire first

Ports on the back of a basic laptop PC. From left to right: Ethernet,.4-pin FireWire, USB 2.0 sockets and analogue video output.

appeared in the mainstream world with DV camcorders, long before any video editing systems supported it. These connections allow data to be copied direct from the camcorder or digital tape in its native digital format without any need for decoding and re-encoding, thereby preserving quality and creating a digital clone of the original recording. This single feature is one of the main reasons that DV became such a revolutionary format, in that it massively reduced the problems of generation loss that plagued analogue editing when footage was copied from tape to tape. Maddeningly, outdated import regulations in the European Union dictate that video recorders (or any AV device capable of receiving an external video signal) were liable to more import duty than cameras with no video inputs. In the early days of DV the result was that most camcorders had their DV inputs disabled when they were imported into the EU. Things are much better today, but many of the cheapest budget camcorders are still being

'nEUtered', so be warned. There are two types of FireWire connection, however. A small, four-pin socket is found on digital camcorders, while a larger, six-pin socket can be found on most computer systems and some external hard drives. Six-pin ports carry power to devices, while smaller four-pin sockets are unpowered.

PS2 PS2 ports are used to accommodate mice and keyboards on Windows systems, but have been dropped from the specification of Apple Macs in favour of USB. USB keyboards and mice can also be used with Windows PCs, but many editors still prefer to use PS2 devices.

Serial and parallel ports These were once the main interfaces used to connect printers and scanners to computers. Today, they are largely unused since modern peripherals use USB or FireWire connections. These sockets are still available on most Windows PCs but have long since been removed from Macs.

FireWire, USB and headphone sockets are sometimes available on the front of desktop computers, as with this Apple Mac.

FireWire sockets are easily added to computers that do not already have them built into the motherboard. The hardware is already supported by modern Windows operating systems, making installation easy.

Ethernet Ethernet ports are used for networking, allowing two or more computers to share files, and even some peripherals, such as printers. There are currently two popular speeds of Ethernet connection: 100Mbits per second and gigabit connections, allowing data rates of up to 1,000Mbits per second. The latter, high-speed connections are theoretically sufficient to enable video to be edited across several systems, but it is still recommended that you should avoid editing across a network if possible, keeping all media files on the local system.

Modem Traditional modems connect the computer to a phone line, providing access to the internet and also allowing the system to be used as a fax machine if you wish. Today's high-speed broadband connections, however, have computers connected to the internet via USB or Ethernet ports and so traditional phone modems are becoming less essential for mainstream computer users.

Video out Many computer video cards offer analogue video output, for sending signals to a TV set. These often come in the form of S-video sockets which can then be made to feed a composite signal with the correct adapter. Some media software programs now use video outputs to provide a high quality output for viewing edits in progress or DVD menus.

Monitor A monitor is the computer's visual display – showing the media with which you are working, as well as a visual interface for the software itself. Until recently, monitors were large cathode ray tubes, similar to traditional CRT TV sets, but with a higher resolution and no tuner for receiving off-air broadcast. As such, they were also heavy and took up a lot of desk space. Today, flat-panel LCD and TFT displays are becoming the norm, thanks to falling prices. These panels take up much less space, are less prone to flicker and produce less heat than their CRT counterparts.

Mouse A mouse is a pointing device, moving a small cursor around the screen and allowing users to select, edit and move objects. A single click will select an item, while double-clicks start programs and run media files. Holding a mouse button down when moving the mouse drags objects around the screen. For many years now, Windows PCs have used mice with two buttons – the left-hand side button makes normal selections, while the right one calls up context-sensitive menus with options that relate only to the item beneath. More advanced is the wheel mouse, which features a small wheel between the two buttons, used for scrolling up and down through long web pages

A large, high-quality monitor is essential for almost all media work. TFT monitors take up less space than CRT models, are less prone to flicker and generate less heat.

At least two buttons and a scroll wheel are useful when choosing a mouse.

function keys for making software commands. Some keyboards also have additional buttons for launching commonly used applications or for easily navigating internet browsers. Keyboards are also hugely useful when editing video. Most advanced editing programs assign keyboard strokes to editing commands, making many tasks much quicker than the use of the mouse to browse menus or click buttons. Some hardware companies even produce special keyboards designed for use with certain editing programs. As well as alphanumeric markings on keys, they also feature icons to indicate their functions when used with that application. Keys on these keyboards are often colour-coded too, to help users to identify and find them quickly.

or text documents. This wheel is also used by some editing programs for advancing forward and back frame-by-frame through video clips. Even more advanced mice feature five or more buttons, providing quick shortcuts for common commands, such as moving forwards and back between web pages.

Keyboard Keyboards are used to enter data into computers and have keys arranged in the same manner as on a traditional typewriter, but with the addition of computer-specific keys such as page scrolling, navigation and

The Operating System

Operating systems are programs that provide a working environment in which other programs can exist and function. The most common operating systems in mainstream computing are Microsoft Windows and Apple's Mac OS. It is this operating system that greets you when the computer is first booted up and provides easy access to files, folders and programs. Operating systems are not normally

To master keyboard shortcuts is very worthwhile since it makes the whole editing process much quicker. Some companies even provide colour-coded keyboards especially for use with specific programs.

interchangeable across computer types – although Apple's decision to use Intel processors in new Macs means that they can be used with Windows or MacOS – and computer programs are generally written for only one platform or the other. Programs such as Adobe Photoshop or After Effects come in a Mac or Windows option, but users will have to choose one or the other, there's no one program that installs freely on either operating system.

Advanced hardware

Any new computer with a FireWire connection and sufficient hard drive capacity should be up to the job of editing DV video. And the latest generation of dual-core processor systems will have little trouble in editing high-definition HDV. There are still many advanced hardware add-ons available, however. For the most part, they serve two purposes: to provide immediate playback of complex effects without your having to spend time rendering them first, and to provide the computer with analogue video inputs and outputs (FireWire can carry only digital signals). Real-time accelerator boards normally double as analogue capture devices, but can be expensive and restrictive, working only with one specific editing program. There is still a good case for investing, however, if you're prone to using many visual effects and getting a lot of paying video work. You should really consider real-time hardware only if time is money and if you are getting enough paying work to justify the expense.

Editing Software

Dedicated software is needed to edit video. If you are buying a Mac, it will already come with Apple's basic editor iMovie installed; iMovie is a very capable editor for occasional home use, but lacks the sophistication and control you will need for more advanced projects and paying work. Microsoft Windows operating systems now come with an editor too, a little program called Moviemaker, but it is little more than a taster to whet your appetite and drive you out to buy something better. On the Mac platform, your choice of advanced editors is largely limited to Apple's own Final Cut Express and Final Cut Pro, both very well featured applications, but rather expensive for the novice and representing a

Real-time accelerators such as the one pictured here by Matrox, are expensive, but make a worthwhile investment for jobbing video makers who actually make money from their work.

*Control surfaces,
such as this one
provide a physical,
tactile environment
for performing
tasks such as sound
mixing. You can
also get controllers
that simulate the
jog/shuttle dials of
a linear editing
console.*

very steep learning curve for anyone moving up from iMovie. The choice is much greater for Windows users and fierce competition also means keener prices. Software developers such as Pinnacle and Ulead offer some first-rate editing tools for beginners and there is a lot of choice for advanced users and professionals from the likes of Adobe, Avid, Canopus, Ulead and Sony. As with most things, don't feel that you absolutely must have the most expensive or advanced programs, think carefully about the kind of videos you want to make and how you plan to go about making them. Then buy

what you need and what is most appropriate for the projects you have planned. By the end of this book I hope to have given you a good idea of how programs differ and how they are suited to different tasks and approaches to editing.

System Stability

Of all the problems that video editors are likely to face, system instability and computer crashes are by far the most common. Your workflow is much more likely to be hampered by gremlins in the hardware or software bugs

Getting Comfy – Being Practical Isn't Enough

You're going to spend many hours at a computer, editing video, so it pays to make it as comfortable a working environment as possible. Go for the fastest computer you can justify buying to avoid unnecessary waiting around for data to crunch. Also, go for large monitors, preferably two, so that the desktop and editing software can be spread across them. If you're

editing DV, a small portable TV set or video monitor is a bonus too, providing you with a constant, full-screen view of the movie you are working on. At the time of writing, this can be achieved with HDV only if you invest in expensive real-time hardware. Good speakers are also a necessity – go for surround-sound speakers if you plan to mix 5.1 for DVD.

Buying or Building a System

If you're going the Mac route you have no choice but to buy a computer off the shelf. There are some specialist retailers that will offer a dedicated video editing Mac, but the most common approach is to buy one of Apple's ready-made Macs, as listed on the company's website. With iMovie and iDVD already installed, beginners can start using it straight out of the box, while more advanced editors will need to spend extra on Final Cut Express or Final Cut Pro, as well as DVD Studio Pro if they have need of high-end DVD authoring tools. The only immediate weaknesses you are likely to find in a new, off-the-shelf Mac are its memory and hard drive capacity, both of which can be upgraded easily and at little cost. The world of Windows PCs is much more diverse, offering you three routes: buying an off-the-shelf computer from a high-street retailer and upgrading it for use as a video editor; buying a dedicated, purpose-built video editing system, or building your own from scratch. If you're serious about video editing as a profession, it can make good sense to buy a purpose-built machine. Find a good, reputable company that builds nothing but video editing systems, they often advertise in DV-related magazines. These system builders have spent time working with different combinations of hardware and software to identify (and ultimately avoid) any potential conflicts that might crop up. You will also get excellent technical support, which is a blessing for professionals for whom time is money and a crashed system means a loss of earnings. If, on the other hand, you choose to build or adapt your own computer, keep a keen eye on the system requirements of any video software or hardware you plan to use, and spend time online, browsing self-help messageboards for the kit you plan to use. It doesn't take long for users to encounter bugs and conflicts if there are any to be found.

than by any learning curve you choose to tackle. If you are editing video for a living and can't afford too much downtime, your best bet is to invest a little extra money in a purpose-built editing computer from a specialized system builder. Regardless of how you get your computer, however, your best move for keeping it gremlin-free is to limit its usage purely to video editing. Don't use it for office work, gaming, day-to-day web surfing or emails. And on no account should you install any of those trial programs that come with monthly computer magazines, they just add fat to your system, overcomplicate things and slow down the applications that are already there. Regardless of what you may read or hear, Mac and Windows computers are equally as likely to crash. But, in either case, the key to stability and stress-free editing is to keep the job descriptions simple and the hard drives clear of unnecessary junk. Don't treat them as jacks-of-all-trades since performance will be compromised in every department.

3 SHOOTING FOR THE EDIT

A simple rule: garbage in, garbage out... There are two expressions that have become a cliché of the video-editing world. One is 'fix it in post', the other being 'garbage in, garbage out'. The first is a cry of resignation from frustrated video makers unable to get the shots they want or the sound they need. All hope is placed on the video editor and sound mixer (post-production crew) to rescue their bad work and turn poor source material into a worthwhile, watchable movie. Garbage in..., on the other hand, is the voice of realism that comes from post-production workers. Put simply, if you give an editor garbage to edit, you can only hope for neatly trimmed garbage in return.

A video editor can find bad footage and cut it from a movie, creatively cutting sound to smooth over any gaps or jumps where the footage had to be truncated, and a sound editor can recreate an entire soundtrack from scratch, including the replacing of dialogue, spot effects, footsteps and background noise. All these processes take time, however, and cost a lot of money in the real world of film and TV production. Due care and attention at the shooting stage will help to avoid this drain on time and money, allow the editors to edit the movie creatively rather than to plaster cracks and ultimately give the director a finished product that closely resembles his or her original vision. Give good editors a rough diamond, and they'll reward you with a precious gem.

SHOOTING FOR THE EDIT – IMPORTANT CONSIDERATIONS

My previous book, *Digital Camcorder*, covered digital video shooting techniques with an eye to gathering useful, edit-friendly footage. There is more in that book than could comfortably fit into one chapter of an editing text, but here are some basic pointers to keep in mind: The key to shooting strong narrative video is to have a clear idea in your head of how your finished movie is going to look. Know how much coverage you need of specific scenes and situations, and remember to think clearly about continuity. In particular, the way in which you position the camera so as not to overstep the '180-degree line', an imaginary line that could be either an eyeline or line of motion (switching sides from one shot to the next would give the impression of a change of direction). Keep shots varied, recording a good mix of wide establishing shots, medium shots, close-ups and illustrative details (or 'cutaways'), which can be used for emphasis or as emergency plasters if need be. Also try to avoid playing with digital video effects on your camcorder. All of these tricks and gadgets can be applied at the editing stage and, unlike the camcorder's effects, you can remove them if you don't like them. Be creative with your use of lights and reflection though and take an interest in some of the more subtle lens filters that are available, such as neutral density filters for cutting down incoming light in very bright environments or circular polarizing filters which remove reflections from glass and water, as well as boosting contrast and making skies more dramatic. Always remember that, with a good editorial brain in your head, your movie will not suffer from shooting too much video, but you might find yourself struggling at the edit stage if you've shot too little.

SOUND – LIVE SOUND AND WILD TRACKS

Sound is half of the picture – that's another popular motto in the video production world, and an accurate one. A good soundtrack can persuade viewers to overlook mundane visuals, while badly recorded or badly mixed sound will serve to amplify the effect of visual flaws. In short, good sound can make your movie look good too. Your first step when it comes to sound design is to invest in a good external microphone for your camcorder (providing the camcorder will accommodate one). The built-in microphones on camcorders are never suitable for serious sound recording. They're normally omnidirectional, picking up sound from all directions and cannot be brought in close to subjects without pulling in the camera too. Different types of mic are suited to different jobs, so be prepared to amass a small collection as you move from one project to the next. Highly directional shotgun mics are often ideal for dramas, while tie-clip microphones and interview mics can be more convenient for documentary, interviews, and presenter-led projects.

As well as recording clean speech, background noise and spot effects have to be covered. Recording 'wild tracks', consisting of a minute or two of uninterrupted background ambience, can be a huge help in the editing stage, as can clear recordings of key object sounds such as clocks, footsteps and machinery. Granted, some of these sound effects are available in specialized CDs and movie scoring software, but it can also be safer and better to record them yourself and take full control over the tone of your movie.

Directional shotgun microphones are ideal if you're shooting solo and don't want a mic in shot.

Interview microphones are fine for many documentary situations where there is no need to hide the presence of a camcorder. They are useful for bringing the microphone up close to the subject.

A tie-clip microphone also allows close-up micing, but is slightly more discreet. It also frees up the subject's hands.

LOGS AND LISTS – GATHER ALL PAPERWORK

Keeping notes during a video shoot is often seen as a chore, most video makers like to get ahead, keep the momentum going and avoid the hassle of note-taking. The truth is though, that a comprehensive set of notes can be a monumental time saver at the edit stage. Listing all the shots on a tape, their duration (or, at least, the start and end timecodes), and comments, even if they are just as basic as 'good' or 'bad', will save the hassle of capturing hours of unnecessary material on the computer and watching it all through to make logs after the fact. If you know that some pieces of footage are not worth bothering with you can ignore them completely and concentrate more on the good stuff. Notes will also make sure that you don't miss useful close-ups and cutaways that might have been recorded at a different time. Even though it is likely that you will be editing your own footage, don't trick yourself into apathy with assurances that you will remember exactly what you've got and where it is. You would be amazed how easy it is to 'lose' essential media later.

4 LIBRARIANSHIP AND VIDEO CAPTURE

Video capture is the process by which media is brought into the computer and stored on its hard drives ready for editing. Video files are very data-heavy and, even with today's colossally huge hard drives, it can be easy to run out of space. Consider that one hour of DV footage will require about 13GB (gigabytes) of hard drive space. HDV video in its native format will require about 11GB, but many high-definition video editors require footage to be transcoded into an I-frame format for editing, which can occupy up to seven times more space than the original MPEG footage. And as well as needing hard drive space for video capture, you will also need to keep space free for rendering. All transition effects, colour correction effects, video filters and titles will need to be remade as new video files before being ready for output. And if you are editing HDV, the movie will almost certainly need to be recreated in full as a self-contained video file before it can be recorded back to tape.

Hard drive capacities can be quite enormous these days, but then operating systems and software have become more code-heavy and demanding on space too. Even if your computer system has a very big hard drive for the main system, it is still advisable to invest in a second drive to be used purely for media files. Many video editors take things further and buy a third for storage of rendered files. Additional drives are easily added to most desktop computers, but you don't have that kind of freedom when it comes to laptops or small, self-contained computers such as the Apple iMac. In that case, an external drive – connected either by FireWire or USB 2.0 – will do the trick for DV editing, although it might prove a little slow for some high-definition HDV applications.

GETTING PREPARED

It is a fair assumption that anyone taking the plunge into the DV editing world will have a reasonably modern PC or Mac computer. On the Mac side, that means that you are running some version of Mac OS X or later, and for Windows, that it is at least Windows XP or Vista. If your operating system is any older, think carefully about upgrading your setup before moving any further. Of course, video editing was being done successfully long before Mac OS X or Windows XP appeared on the scene, but it was never as easy to set up a working system as it is today.

If you need to install a new hard drive for capture, take a good look at the computer's motherboard to see the types of hard drive that it will support – typically ATA or SATA drives. SATA drives are easy to install and can often even take power and data through a single cable. Older ATA drives require a little more thought, especially when it comes to identifying them as 'master' or 'slave' devices, depending on how they are connected to the system. Formatting hard drives for Mac computers is a simple, quick and utterly painless procedure. And, while it is hardly an intelligence test for Windows users, there is considerably more waiting time involved and a few more hoops to jump through. For Windows systems, a new hard drive must first be partitioned, meaning that some or all of its capacity must be assigned for use as a single drive. You can opt to have two or more partitions on a drive, but, in most cases, one partition encompassing its full capacity is most appropriate. Next the drive must be formatted so that it is primed for reading and writing. On Macs, this takes only a few seconds, while Windows machines take a very long time over it, and the higher the drive's capacity, the longer formatting will take. Windows users should also be aware of the different formatting options available. In particular, it is important *not* to choose FAT file systems since this will impose 2 or 4GB file size restrictions on all data, which is a huge inconvenience when working with big video files. Instead, make sure to select NTFS formatting.

How will you connect your camcorder or video deck to the computer? If you're working with DV or HDV video, a single FireWire cable should be all you need, provided that the machine has a FireWire socket; all modern Macs do, but many Windows PCs still do not. If you are capturing from an older analogue camcorder or an analogue VCR you will need either a converter to change an analogue signal into DV and feed it to the machine via FireWire, or – as a less costly solution – an analogue capture device such as a TV tuner. In either case, be sure that the capture hardware is correctly installed and recognized by the editing software you have chosen to use. It used to be the case that analogue capture devices would work only with the software they came bundled with, but continued

---

Check System Requirements

Before buying, installing and using any piece of media software, read up on its minimum and recommended system requirements to ensure that your computer is up to the job of running it. The consideration is not just the processor speed, available RAM or your choice of graphics card – Adobe's Premiere Pro and Premiere Elements programs, for example, tend to require processor extension sets that no longer exist on many older systems (nor more recent machines with AMD processors).

---

standardization – mainly among consumer-level TV tuners – has lead to greater compatibility across a large range of products. There is a much larger choice of TV tuners and analogue capture devices for Windows systems than there is for Macs, but affordable options are available on both platforms if you ask a specialized dealer.

The obvious advantage with digital video transfer via FireWire is that there is no need to convert the video signal into another format. While analogue devices take decompressed analogue signals and convert them into digital formats, digital capture simply copies ones and zeroes from the source tape, creating a clone of the original footage on the system's hard drive. At least that's the case for DV editing and some HDV-compliant setups. Some programs, however, still insist on converting HDV footage to a more data-heavy I-frame format for editing, and, while this has some advantages over cutting IPB-frame MPEG-2, it does include additional steps of decoding and encoding, which DV and native HDV editors neatly sidestep.

We take a very obvious step away from

Analogue to DV converters such as this are useful for feeding analogue video and audio signals to a PC via FireWire and for sending DV video back out to analogue tape from the computer without the need for extra internal hardware.

traditional analogue and digital capture methods when acquiring footage from DVD or hard-drive-based camcorders. In these cases there is no need to play back media in real time, as it would be if read from a tape. Different models of camcorder have different ways of working, but most connect to computers via USB 2.0 channels, and the process of video capture is close to that of copying data from one drive to another, or importing photos from a digital stills camera. For more on this see the section below in this chapter on preparing and importing media. Similar methods will also apply to the acquisition of video from mobile phones, and, while this footage isn't ideal for your home movies, people still shoot with their phones and an increasing number of consumer-level editing programs now support the common mobile video formats.

On an immediate and practical level, it is useful to have your source tapes carefully labelled and to have a fair idea of what each one contains. Comprehensive notes detailing the content and identifying good and bad 'takes' (versions of a particular shot) are also useful and form the backbone of administration on commercial film and video shoots. If necessary, watch the tapes through again after the shoot and, before you start editing, take notes along the way to ensure

that you know what you will need for your movie and what you won't. Here's where you will need to balance the benefit of content against time and practicality. Shooting lots of additional material such as establishing shots and detailed cutaways is good practice, but bringing *everything* into the system for editing without first planning out the edit itself can leave you indecisive and slow the editing process down to a crawl. If you are unsure about whether certain shots will be needed it is obviously better to have them handy than not, but it can also help to capture them into separate 'bins' (if your software supports it) to keep your workspace organized. See the section on bins, racks and sequences near the end of this chapter.

DEVICE CONTROL, METADATA AND SCENE DETECTION

As well as the promise of direct digital copying, the capture of footage from digital tape via FireWire has the added advantage of enabling device control from the computer itself. This means that commands such as play, rewind, fast forward, frame advance and record can be made by using the editing software rather than from buttons on the device itself or its remote control. In the old days of expensive linear editing suites, where editing was done by copying footage from one tape to another, accurate device control was essential. And while it is technically not so fundamentally crucial for many non-linear editing projects, the feature is enormously convenient and gives rise to some quite advanced and time-saving tools that we shall see later.

The next advantage of a FireWire link comes with the reading of metadata – simple pieces of information that don't contribute to the picture or sound. In DV recordings, these amount to timecode, date and time. Timecode is the most important element for a video editor to consider since it forms the basis of many editing decisions and can either help or hinder your editing process depending on the level of your understanding and vigilance. In essence, timecode is a number that is assigned to each individual frame of video. It is represented in hours, minutes, seconds and frames, with each unit being separated with a

colon symbol for PAL footage or a semicolon for NTSC. So the first frame on a PAL DV recording would be 00:00:00:01, the second would be 00:00:00:02 and so on. In DV and HDV recordings, as with all professional broadcast formats (be they analogue or digital), timecode assigns a unique number to a specific frame, thus ensuring that no two frames are given the same code on a single tape. That's the plan at least – many camcorder users are in the habit of watching their footage back and leaving a small gap of blank tape between shots, prompting the camcorder to reset its timecode allocation to zero, in the assumption that you have given it a blank tape. This will cause great confusion later as we explore more advanced video capture tools. If you plan to review footage as you go be sure to 'black' your tapes first by running them through the camcorder with a lens cap in place and a headphone plugged into

DV capture programs, such as Ulead's Quick Scan – provided with VideoStudio and MediaStudio Pro – scan tapes at high speed and create logs of each individual shot, based on changes in time and date stamp. Specific scenes can then be selected for batch capture.

the mic socket to give black, silent footage with a continuous timecode.

Date and time information may seem trivial and some camcorder users never bother to enter this data into their camcorders, but it serves a far greater purpose than just to remind you when a shot was taken. Almost all video editing programs are capable of monitoring the video's date and time data during capture. A sudden break in the time stamp indicates that recording was paused and restarted some time later. And this, therefore, indicates the end of one shot and the beginning of another. Scene detection tools in mainstream editing programs provide an easy way to break long video captures down into smaller chunks based on when the camcorder was paused, which is a great help when it comes to organizing large amounts of footage.

TAPE LOGGING AND BATCH CAPTURE

The use of timecode, scene detection tools and device control lend themselves to some nicely advanced video capture tools designed to simplify the acquisition process as well as to allow the careful cataloguing and organization of media from the start. At the entry level, one of the most helpful and intuitive introductions to capture management comes through tape logging applications. Ulead's VideoStudio and MediaStudio editors come with these tools built in, under the name DV QuickScan. Other companies' editors have been slow to catch on, but self-contained tape logging tools are available at a low price if you run a quick search on the internet. The tools will play each tape (either in real time or by using high-speed picture search) and monitor changes in date and time stamp to record the points at which scene changes occur. The program puts together a database of the tape's component clips, featuring a thumbnail image to represent

Other Metadata

Some very detailed information is written to DV tape alongside audio and video. If you monitor footage with the correct software it is possible to find information on the focal length, shutter speed and aperture settings used for each individual frame of footage. Camcorder make and model details are also stored in many cases. This is similar to the metadata that accompanies images from good digital stills cameras, but, as it serves little direct purpose for the job of video editing, it is unlikely that you will see it represented in your editing software.

each clip, along with a name, duration and its timecode information. Names can be changed so that they mean something sensible to the editor, and logged clips can be made to play back from tape by using automated device control commands from the computer. In most cases, logs can be saved for future reference, but, most importantly, they allow users to select which clips will be captured and which ones won't be – the required clips are then captured in an automated batch, with the only required human intervention being to change tapes if or when prompted.

Automated tape scanning and logging tools are relatively new to the entry-level mainstream market, but more detailed, hands-on logging methods have been the mainstay for professional and prosumer-level video editing for many years. Device control tools are available at this level, but it is much more common to have users manually select the start and the end point of each required clip based on its timecode. The logged clips are then added to a list or created in the editor's project windows as an 'offline clip' which contains basic information but no actual audio or video media. As we have already seen with automated tape scanning tools, several clips can be selected for *batch capture*, being brought into the system in a single, unattended operation, with the computer controlling camcorder playback and involving the user only when tapes need to be changed.

PREPARING AND IMPORTING MEDIA – VIDEO, SOUND AND GRAPHICS

Not all the media you need for a video edit will come from video tape. Many camcorder formats are now tape-free, using DVD discs, hard drives and data cards instead of traditional cassettes. Furthermore, an increasing number of people are shooting video on their mobile phones. At the time of writing, mobile phone footage is of a very low quality compared with what you will get from a dedicated camcorder, but that does not seem to stop them from using their phones to capture precious moments, and it is only natural that they'll want to include this footage in video edits. Furthermore, there is also the need to use still photos, graphics such as maps and corporate logos, wild tracks and music in video productions. All these elements need to be brought on to the computer and presented to the editing software in a format that it can understand. The more recent a video editing program is the more likely it is to be compatible with new multimedia formats from mobile phones, as well as the various MPEG-2 and MPEG-4 files created by DVD, hard drive and memory card-based camcorders. These formats are also more likely to be supported at the entry level than in pro software, where it is assumed that movie makers will be using established, high-quality video formats for their shoots.

A quick glance at the program's manual will tell you exactly which file formats and video or audio codecs are supported by a particular program. Anything falling outside this list is likely to be rejected. This can be more of a hassle for video and audio files (particularly those downloaded from the net or created with mobile phones or digital stills cameras) than it is for images. That said, some picture editors have their own proprietary formats for layered or object-based images. Photoshop's PSD file format is widely supported by programs at the prosumer level but, by contrast, Ulead's PhotoImpact program creates a UFO file for its layered images, which is not understood nor accepted by any programs other than Ulead's own.

For video, it is good practice to keep all footage in the same format and created to the

The Benefit of Offline Clips

Many editors at the professional and prosumer level support 'offline clips'. These are video clips that have been logged on a tape but not yet captured to the editing computer. These clips can be added to a project timeline and generally identified as a video clip, but they can't be played since they contain no actual video or audio information. Before they can be used properly, the editing software must capture the media from tape. The advantage of having offline clips as opposed to a simple batch capture list is that media can be 'offlined' at the end of a job, meaning that all editing information remains in the project, but the media itself is erased. Doing this frees up hard drive space for other work, but provides the ability to recreate the project later by automatically recapturing the required footage if it needs to be revisited or revised.

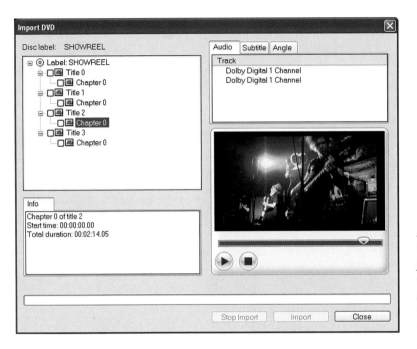

More and more video editing programs allow video to be ripped direct from unprotected DVD video discs.

same specification. So if the bulk of your footage is DV, PAL standard and 16:9 widescreen, all video material from other sources should be prepared or converted to the same specification. Many editors will claim to

Search Tools

A growing number of video editing programs are being given search tools by which users can hunt down the clips they need based on the name, tape, date or clip notes that could have been added at the capture stage. Such tools are a great asset but are useful only if you have taken the time in the first place to set the date and time on the camcorder, label tapes, name clips and to enter detailed notes. Without this information, you're still flying blind.

support different formats and specifications in a single edit, but the speed and the quality of playback can suffer greatly as a result. Similarly, many editors support highly compressed audio formats such as MP3 or Microsoft's WMA, but the process of decompressing and recompressing these files just burdens the computer with even more work. For the sake of speed and efficiency, it is advisable to prepare all media as uncompressed PCM stereo files in WAV or AIFF format, and with a 16-bit sample size and a sample rate of 48kHz. More on audio and sound editing can be found below.

Still images and graphics should be prepared with a resolution larger than that of the video frame. Regardless of whether your video is 4:3 or 16:9 widescreen, DV video has an image resolution of 720×576 pixels (for PAL footage) or 720×480 pixels (for NTSC). HDV frame sizes are 1440×1080 pixels or 1280×720 pixels, irrespective of frame rate or territory. DV and 1080i HDV frames are stretched by the playback software to take the required shape. Images prepared at a smaller size will have to be magnified in order to fill the video frame, resulting in some noticeable loss of clarity. If pictures are made to a higher resolution, however, they will have to be scaled down to fit the frame and will therefore remain sharp and clear. They also have the advantage of allowing you to zoom into images or pan across them without any apparent loss of picture quality. For more on common format specifications, see the Glossary.

Once created to the appropriate specification, files can be imported into the same folders and bins as footage that has been captured from tape. Video shot on hard drive or DVD-based camcorders is often imported in this manner too – copied as data files from the source device and imported into the software. While there are definite pros and cons to using these camcorders, there is a very obvious advantage in being able to get footage on to editing computers at a much faster speed than with tapes, where video must be played back in real time.

ADVANCED SOFTWARE – BINS, RACKS AND SEQUENCES

Organizing media files can be a mammoth challenge, large projects normally require hundreds of individual shots, plus illustrative photos, graphics and audio tracks to provide music, ambience, narration and spot effects. Keeping on top of all this material is nigh impossible if it is all crammed into a single folder. In many entry-level editors there is not much you can do to organize these clips, but more advanced editors, such as Adobe Premiere Pro and Avid Liquid, allow numerous folders (known as 'bins') to be created and named so that clips can be sorted according to file type, subject, which scene they belong to or

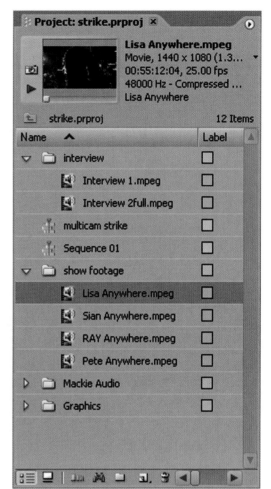

Bins, folders or racks are simply directories within the software interface in which media files are organized. These folders may not actually exist on the system's hard drives, they just provide a virtual directory system in which materials are organized in the editing environment.

by any criterion you wish. Taking the time to organize media in this manner will save a lot of time later in the editing process as you attempt to track down specific shots or sound effects.

5   SIMPLE STRUCTURE

As a video editor, your primary task is to present video, audio and images as a narrative sequence, normally a comprehensible one that clearly conveys a point, but this isn't always a requirement. Editing is a process of trimming unwanted material from clips, arranging them into a narrative sequence, establishing connections between images and ideas and evoking a mood with the use of juxtaposition, pacing and rhythm. These tasks are achieved most often with simple, fuss-free editing tools rather than with the crazy effects and filters that dominate each software developer's sales patter. Organizing your ideas, creating a seamless chain of thoughts and pacing the stream of information to yield the maximum emotive response from your viewer is a matter of good judgement and careful cutting, and they are factors that no effects package can help you with.

EDITING BASICS – ASSEMBLE, INSERT AND SPLIT EDITS

At the most basic level, there are three types of edit: the assemble edit, insert edit and split edit. To understand how each differs from the others it is important to think of video and audio as two media elements that can be manipulated independently of one another.

Assemble Editing

This is possibly the most basic form of video editing. Clips are lined up end-to-end, with video and audio being cut at exactly the same moment. The approach is similar to crude linear editing between two VHS video recorders, as footage is simply transferred from tape to tape with no separation of picture and sound.

Look and Listen

Next time you watch a movie or TV programme try to spot the number of times the editor makes use of cutaways and split edits as opposed to straightforward assemble edits. Insert edits are used extensively for illustrative purposes, bringing in illustrations, graphics and cutaway details without disturbing the flow of the video.

Split edits, on the other hand, are the mainstay of dialogue and interview videos and are also used as a neat device to lead viewers from one scene to another without the need for distracting transitional effects or scene-setting shots that could otherwise slow down the pace.

Standard assemble editing as seen on the storyboard of a beginner's editing program, in this case, Apple's iMovie. Notice that the storyboard represents each clip's thumbnail image as the same size, regardless of its duration. There is also no immediate representation of the movie's soundtrack. But the most important thing to notice about assemble editing is that picture and sound are always cut at exactly the same moment.

Insert Editing

This begins to take a more controlled approach to video. In this case, video is cut into an existing sequence in such a way that only the picture is replaced and the existing audio is left intact. As an example, imagine a TV news interview about seal migration in Shetland, an insert edit would allow us to cut the picture away from the speaker to footage of the seals themselves without having to disturb the spoken soundtrack. Depending on the nature of the edit, sound from the inserted clip can be mixed in with the existing audio or removed altogether.

Split Edits

These take a very subtle approach as picture and sound are cut at slightly different times. They are often referred to as L-cuts and J-cuts because of the shapes they form on an editor's timeline. In an L-cut the picture is cut before sound and in a J-cut sound is cut before the picture. The technique is invaluable in creating seamless edits and allowing a sequence to flow naturally. An obvious example of where split edits come into their own is in the editing of dialogue or interview footage recorded in several takes with a single camera doing separate shots of each speaker. Split edits will allow you to show a character's facial

ABOVE: *Insert edits – as with all editing techniques – are better illustrated on a timeline interface. Picture and sound tracks are clearly visible and can be independently edited.*

BELOW: *An insert edit in Pinnacle Studio. Notice that the inserted clip is placed on a new timeline track, intended to allow picture-in-picture effects.*

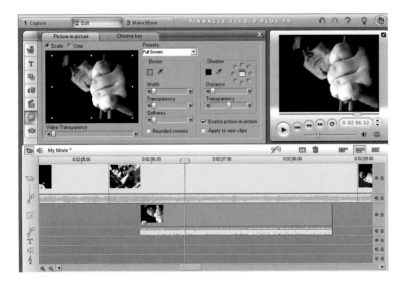

In more advanced editors – such as Adobe Premiere Pro, pictured here – the user is given access to an unlimited number of video tracks on the timeline. Insert editing is easily achieved by just dropping the new clip on to a new track and adjusting its audio to suit.

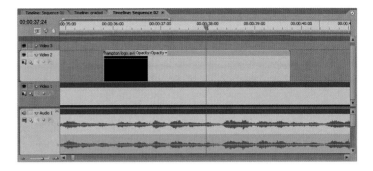

reactions before he starts speaking, or to have him interrupt the flow of things by bringing his voice in before cutting to his image. This simple technique is immensely powerful in persuading viewers that events are happening in real time, even if all the shots were gathered over several hours or even days.

RIGHT: *Dedicated mark-in and -out commands for video and audio elements of a clip, as seen in Adobe Premiere Pro.*

The basic shape of an L-cut on Pinnacle Studio's timeline...

...and a J-cut.

BASICS APPLIED – STORYBOARDS AND TIMELINES

There are two common interfaces in which video edits are compiled: storyboards and timelines. Storyboards are common in budget entry-level editing programs and are designed to make the task of assembling video sequences as simple as possible. Timelines are offered by all good editors and are typically the only style of interface that you will find in prosumer and professional applications.

In a storyboard each clip used in an edit is represented by a single thumbnail image showing its first frame. These thumbnails are all of the same size, regardless of the clip's duration, and represent its video and audio elements, meaning that picture and sound are always cut at the same time. There is no space on a storyboard for additional video or extra audio channels, which greatly limits the complexity and control of your edit. As you can see from the accompanying image, storyboards offer a straightforward view of the sequence of clips, but provide no visual reference to pacing or the interplay between visuals and audio. If you are working entirely in a storyboard-based environment it is unlikely that you will be able to achieve anything more sophisticated than a simple assemble edit.

Timelines are much more complex in structure but give the editor far more freedom than storyboards. In many ways they are easier to use because of it. A timeline presents the user with several media tracks on to which video, audio or titles can be placed. These tracks are set against a time scale, giving an immediate representation of the relative duration of each clip. As can be seen in this image, more than one video and audio element can be applied simultaneously and there is no need for a clip's picture and sound to be cut at exactly the same moment. While all good editing applications provide a timeline interface, some at the entry level are extremely limited, supporting only a small number of video and audio tracks. Some of the most basic budget editors do not even allow for the independent cutting of a clip's video and audio elements. Timelines in more advanced programs are far more versatile though, with most supporting an unlimited number of tracks and allowing a great amount of control. Most advanced programs even support multiple timelines in a single project, allowing one to be 'nested' inside another, where it is treated as a self-contained clip. The advantage of doing this is that big projects can be edited in smaller chunks before being assembled into a longer movie.

Choose Your Editor Carefully

Most video editing programs now support split editing, but some offerings at the entry level still do not, or, at least, they make it more difficult than it really should be. For a long time, software developers claimed that split editing was a 'pro' feature and too complicated for inclusion in beginners' software. Fortunately, most have now changed their tune, but there are still a few applications available that might leave you feeling stung. Even if a program has a timeline interface, that in itself does not guarantee that you will be able to cut a clip's sound and picture independently. If in doubt, ask on the program's online support forums before spending your money. Remember though, that this limitation only affects a small number of budget programs at the entry-level.

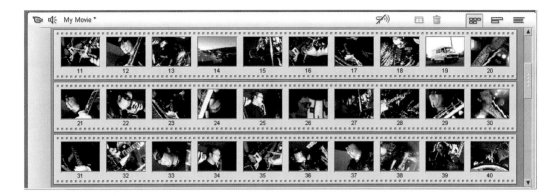

Pinnacle Studio's storyboard interface...

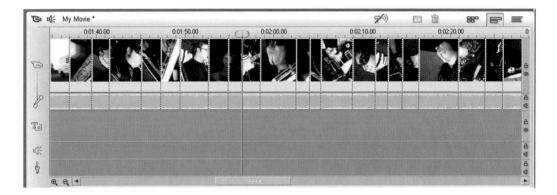

...and its timeline.

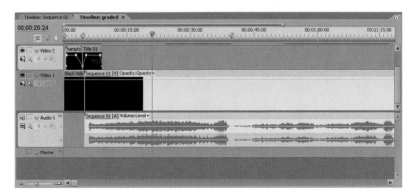

Nested timelines in Adobe Premiere Pro.

THREE-POINT EDITING: THE SOURCE/EDIT INTERFACE

All video editing programs differ in their workflow, but most advanced applications share the same basic approach to simple cutting, using a source/edit environment. The software is laid out with two video monitors: a 'source' monitor in which raw footage is viewed and trimmed of unwanted tops and tails, and an 'edit' monitor which shows the actual edit in progress. Clips are trimmed in the source monitor then sent to the timeline, where they can be viewed as part of the overall cut in the edit monitor. This approach is intended to simulate the common workflow of old linear editing suites and has remained in the non-linear world, because it is a sensible and tactile way to work. Some editing programs, such as Sony's Vegas, take a different approach to editing, but for the most part, a source/edit interface is the norm for 'serious' software.

Clips are loaded into the source monitor by dragging them across from a clip bin or simply by double-clicking on them. The monitor is used to play them through, using dedicated playback buttons, or users can quickly scrub through footage by dragging left or right on a navigation bar directly beneath the video image. Avid's Liquid even allows users to click and drag the mouse left and right on the actual video image. Once you have found the frame at which you want the clip to start this is identified with a 'mark in' command – made by clicking an appropriate button on the source monitor, selecting 'Mark-in' from the program's drop-down menus or hitting the appropriate keys on the keyboard (often the letter 'i'). For more advanced split edits, programs such as Adobe Premiere Pro and Apple's Final Cut Express and Final Cut Pro provide 'mark video-in' and 'mark audio-in' commands. A similar approach is taken in identifying the frame that will end the clip. Play or scrub in the source monitor until that frame is found and select a 'mark-out' command from the monitor's buttons, software menu, or by using a keyboard shortcut (usually 'o'). As before, dedicated 'video-out' and 'audio-out' markers are provided in Premiere Pro and Final Cut.

The third point in our three-point editing process is to identify the place on the timeline at which the incoming clip will be placed. This could be at the very end of a sequence that is being roughly assembled or it could be halfway through an existing piece of footage.

Three-point editing options in Apple's Final Cut Pro.

A single-monitor, drag-and-drop interface is the norm for beginners' editing programs, such as Ulead VideoStudio.

A mark-in command is not necessary at this point – just placing the timeline's scrubbing bar at the required point is often sufficient.

There are different ways to send a clip from the source monitor to the timeline. These may be referred to by several names, such as insert, overwrite or film style cutting, but, at the fundamental level, they allow the incoming media either to replace existing footage on the timeline from the position of the scrubbing bar or to move anything to the right of the bar further down the timeline, to resume at the end of the new clip. Different editing programs offer additional variations on these cutting methods, such as Final Cut Pro and Final Cut Express's additional options of automatically applying transition effects to the cut. There will be more on transitions later in the book.

Three-point editing can also be done by placing in and out markers on the timeline, with either an in or an out point marked on the source footage. Taking this approach, the duration of the incoming clip is denoted by the space allocated on the timeline rather than by

the content of the clip itself. Going one step further, applying in and out points to the source clip and the timeline allows programs such as Final Cut Pro to perform a 'fit to fill' edit, whereby the speed of the source footage is altered to fit the required space.

FREEHAND EDITING – THE DRAG-AND-DROP APPROACH

Formal, three-point editing with a source/edit interface isn't necessary for video editing – it is often quicker and easier to drag and drop clips from window to window or direct on to timeline tracks. This tactile method of dragging, dropping and trimming direct on the timeline is favoured by many video editors and is the standard approach offered at the entry-level by most budget editing programs.

Budget editing programs for beginners typically provide only a single video monitor which serves as both source and edit viewer, depending on which part of the interface is currently selected. As such, the three-point

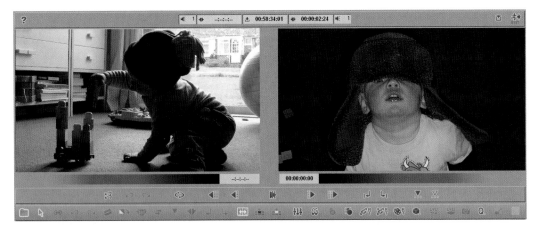

A typical – but very visual and tactile – trimming window, as seen in Avid Liquid.

editing procedure listed above does not apply to these applications. Instead, clips are loaded into the video monitor and trimmed with mark-in and mark-out commands, as we saw previously. That done, they are added to the edit by dragging them from the monitor over to a timeline or storyboard, placing them at the end of a sequence or in between existing clips.

Once on a timeline, a clip can be trimmed further by clicking its edges and dragging left or right to shorten it further or to let it out if it was initially trimmed too tightly. This type of editing technique isn't just restricted to beginners' programs, it is often used with advanced software for those who prefer to take a very tactile, hands-on approach rather than the typical, three-point methods that can often feel mechanical and prescriptive.

TRIMMING WINDOWS

Any decent editor that provides a three-point editing interface will also feature a trimming window for the highly controlled correction of cutting points between clips. They are a common tool among advanced video editors but hardly ever seen at the entry level. A trim window is typically presented as two video monitors, one displaying the last frame of the first clip and the other showing the first frame of the next clip. Accompanying this display is a series of control buttons, allowing cutting points to be changed in one of three ways; first, users can opt to change the mark-out point of the first clip, shuffling the adjacent footage left or right on the timeline to compensate and ensure that no gap is left between them; a second option is to adjust the in-point of the second clip – again, the position of the clip on the timeline moves so as to prevent any gaps forming between them; the third option is to move the cutting point itself, while the video frames maintain their position on the timeline. So as the out-point of clip 1 is let out, the in-point of clip 2 is pulled in. The move is especially useful when trimming between synchronized footage in a multi-angle edit, but is let down in many editors because trimming windows normally require the adjacent clips to be located on the same timeline track.

Trim windows can be operated

quantitatively, using buttons that adjust cutting points in single frames or groups of ten. Alternatively, they can be controlled in a more qualitative manner by clicking and dragging trimming points with the mouse. Either way, the dual-monitor display is updated to show adjacent frames around the cut.

SLIP AND SLIDE EDITING

Each video editor provides its own set of editing tools that can be applied in a click-and-drag manner to the timeline itself. The tools most common to all of them are slip and slide edits. A *slip* edit keeps the clip in a fixed position on the timeline, but slides its in and out points, effectively moving the position of its frames on the timeline. Imagine that you have just laid down a number of shots so that they cut to the beat of a piece of music, but you want to adjust one of them so that certain events coincide with musical cues. The events need to be moved, but the cutting points should stay the same. That's where a slip edit would be used, to select and drag the clip left or right on the

> ### Ripple Edits and Ripple Delete
>
> Ripple editing and ripple deletion are phrases that occur in the majority of editing programs. They refer to the practice of 'closing gaps' on the timeline when clips are trimmed or removed. If, for example, I were to ripple delete a short establishing shot from the beginning of an edit, all clips to the right of it would be pulled to the left, so as to prevent any empty space being left where the clip used to be. Similarly, if I were to trim ten frames from the end of a clip in ripple edit mode, all clips to the right of it would be pulled ten frames to the left.

timeline. There is more on this technique in the 'In the Real World' chapter. By comparison, a *slide* edit moves a clip on the timeline itself, without altering its trimming points; clips to the immediate left or right of it are trimmed to compensate, however.

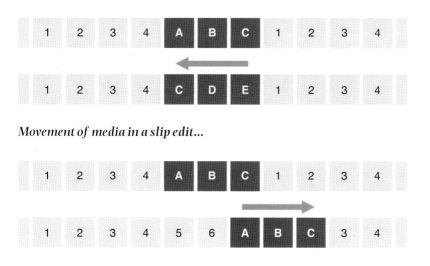

Movement of media in a slip edit...

...and in a slide edit.

6 SOUND EDITING

Sound is every bit as important to a movie as its visuals. A poorly recorded soundtrack will make even the best-looking movie appear weak and amateurish, while slick audio will give your finished product a professional sheen, regardless of how mundane or just plain shoddy the camera work is. The role of sound in drama especially cannot be overstated – some of cinema's most famously scary moments, such as *Psycho*'s shower scene or *Forbidden Planet*'s monster sequences, owe practically everything to sound, be it the tone of recording, the spot effects or the use of music. Many guides to digital video editing will deal with sound quite late in the process, choosing to cover all aspects of visual manipulation and special effects first. I have deliberately chosen to introduce the basics of sound editing earlier in the day to encourage you to take it just as seriously as the main picture cut.

AUDIO LEVELS, RUBBER BANDS, KEYFRAMES AND MIXERS

One of the main tasks in audio editing for video is to correctly control and set levels of every audio element in the mix. Layering audio sources – as you almost certainly will be if you are doing the job properly – has a cumulative effect on the sound volume, and it is easy to make your movie too loud if you're not

Sound Monitoring

If you're going to take sound seriously in your projects you have to be able to hear it properly. Inexpensive computer speakers are often insufficient for the job and will not give you a fair representation of the soundtrack's dynamic range. If you are mixing surround-sound for DVD, I would recommend your investing in a surround-sound amp and some good, solid speakers for close-range monitoring. Even if you are working in stereo, solid hi-fi speakers are a good investment. If space, thin walls, babies or neighbours are a concern though, buy some high quality headphones instead. In addition, try to have your edit routed through to a TV set if possible. Doing this will give you the chance to hear your soundtrack coming from a TV's speakers, allowing you to compare the ideal environment, through your good speakers and headphones with the likely reality of a rattly TV set.

paying attention. Pushing the sound levels too high will lead to distortion and the clipping of the soundtrack, which will sound quite horrible.

Most video editing programs come with a mixer interface, with sliders that simulate the look of a physical mixing console. You will find a slider for the adjustment of sound levels on each audio track on the timeline, as well as another for controlling the 'master' output – sound levels on the overall mix. However you choose to mix each sound element, it is important that the master output levels average at about –6dB and not be allowed to peak above 0dB. Some programs give a numerical scale on their mixers and others don't, but, in the absence of quantitative scales, the output level monitor will be colour-coded, with green being safe levels, yellow approaching the 0dB threshold and red exceeding it. With most editors, adjusting the level slider for a track will reduce or increase its sound level across its entire duration. It can, however, be used to make gradual changes during playback of the movie itself, allowing you to adjust sound levels while watching the action play out.

Controlled changes to any attribute over time, whether the audio levels or visual properties, are brought about with the use of keyframes. These define a specific value for certain properties at a given place on the editor's timeline or frame of footage. If the values for this property differ between two adjacent keyframes, it will be gradually changed over the frames in between. So if the overall movie volume is set to 0dB 2 seconds from the end of the project but dropped to silent on the last frame, the audio levels will steadily drop over those last couple of seconds in order to move smoothly from one value to the next. Depending on the software you use, the progression from one keyframe state to another can be a straight linear one or a more

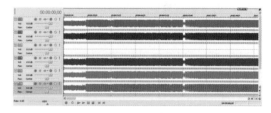

A mixer panel provides virtual sliders representing each audio track on the timeline.

A somewhat different layout is seen in Sony Vegas: each track has its own audio level slider.

Audio monitors are colour-coded to indicate that sound levels are reaching dangerous limits. Try not to let them spend any significant time in the red.

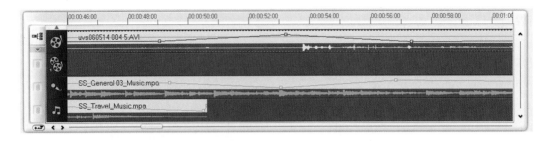

Direct manipulation of rubber bands on the timeline of Ulead VideoStudio.

subtle and curved one, denoting acceleration or deceleration at different times.

When using a mixer panel to alter audio levels during playback keyframes are set on the timeline in real time, and changes are effected accordingly. A more precise method of level adjustment is also possible in most editors by setting keyframes direct on audio tracks and dragging them into position. The lines that represent values between keyframes are known as *rubber bands* and, in the case of audio levels, they are quite easy to understand – a high band means loud, while a low level means quiet. An upwards slant means an increase in volume and a downwards slant denotes a decrease. As well as dragging keyframes into position direct on the timeline, most editors allow their values to be set numerically if need be.

STEREO AND SURROUND-SOUND PANNING

The other set of controls you are likely to find in an audio mixer panel affects panning in stereo space – determining the balance between the left and the right speaker. Panning controls appear as a dial or a slider, each used to shift a sound to the left or to the right. The panning of sound sources helps to create a greater sense of space and depth in your production, giving a general effect of solidity and realism. As with audio levels, panning decisions may be keyframed, giving the illusion of sound sources moving left to right or right to left – useful techniques when introducing spot effects or bringing in 'aside' dialogue from off-screen characters. It is rarely necessary to pan sounds to the extreme left or right in order to create a good stereo effect, just a subtle shift of balance over to one side is normally sufficient to give greater depth to your soundtrack.

The application of keyframes and rubber bands in Premiere Pro.

66

The ever-increasing interest in DVD authoring has also led to surround-sound panning tools being made available in entry-level video editors. In this case, users are faced with six speakers rather than two – left, right and centre speakers at the front, a left- and a right-rear speaker and a subwoofer to provide bass sounds. Regardless of the more complex environment, however, the placement and the panning of sound sources is arguably more intuitive in surround-sound than in stereo, thanks to graphic interfaces in which sources are positioned within a room space (often in relation to the viewer's ideal position in the centre of the room). Entry-level video editors,

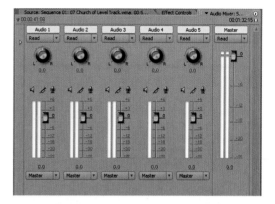

Stereo panning controls often take the form of a dial or a slider.

Visible Waveforms

Audio editing is made altogether more intuitive if your editing software displays visible waveforms on its audio tracks. Waveforms (see below) provide a visual representation of audio content and help users to identify sound cues immediately without them having to rely purely on hearing and lightning reflexes. A good visual waveform will help you to isolate individual words in a speech and will provide an invaluable reference guide for dipping and boosting sound levels when bringing in music or sound effects.

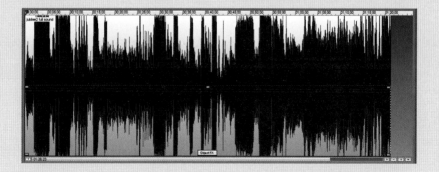

Visible waveforms give an invaluable guide to potential cutting points, allowing you to navigate dialogue and music easily.

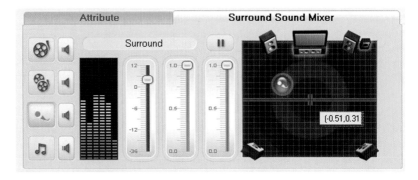

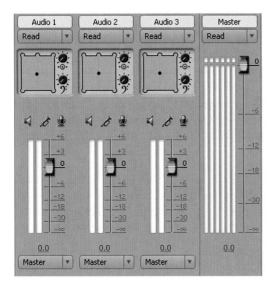

ABOVE: Surround-sound panning at the entry level is often visual, tactile and intuitive.

LEFT: In more advanced software, surround-sound panning becomes more complex and potentially confusing.

BELOW: Sony Vegas takes a different – and welcome – approach to stereo and surround-sound panning, putting the necessary controls at the head of each individual track.

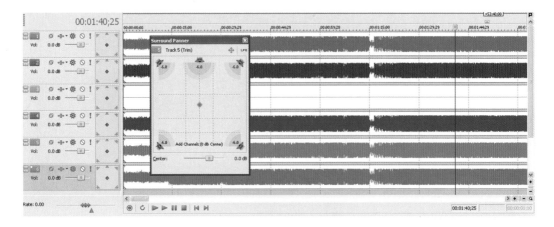

such as Pinnacle Studio, Ulead VideoStudio and Magix MovieEdit, make surround-sound panning easy, even to the point of allowing sounds to be manually moved during playback by clicking and dragging with the mouse. Things become more complicated at the prosumer level, as more quantitative controls are introduced, but programs such as Sony Vegas are still highly graphic and intuitive. At the time of writing, however, other prosumer offerings such as Avid Liquid and Adobe Premiere are considerably more restrictive in their workflow for surround-sound panning, keeping the relevant tools far removed from the main editing interface.

Your decision of whether to mix stereo or surround-sound will ultimately depend on the way in which your movie will be seen. For DVD distribution, surround-sound is a good bet – but only if you are using a moderately advanced authoring application that will support Dolby Surround audio tracks. Beyond this, most tape formats, including VHS, DV and HDV, have no true surround-sound support. And while surround-sound can now be incorporated into streaming video formats such as Windows Media, doing so results in the creation of large files with a very high bandwidth, making them unsuitable for the majority of internet users. Surround-sound should also be considered if you're planning to present movies to film festivals. As these events normally take place in established cinemas, facilities might exist for you to provide separate video and surround-sound tracks for synchronized playback.

AUDIO TRANSITIONS

Transition effects ease the move from one clip to another with a gradual change. Video transitions come in a wild and crazy assortment of flavours, but with audio we are largely restricted to simple crossfades, causing the outgoing clip to fade out to silence as the incoming media fades in. Audio transitions can have a very obvious effect in bridging the move between scenes and locations. A sudden change in the type and tone of a soundtrack can be annoying and distracting, so a short crossfade helps to lessen the impact. On a more subtle level, crossfades are a life-saver when pruning a single video scene. Imagine that we have recorded an interview with two cameras in a location with some audible background chatter, for instance, from a crowd of people, passing traffic or machinery within earshot. The coverage of two cameras gives us the freedom to remove stutters, repetition or just plain boring content from the final edit without any obvious visual jumps. But sudden changes in the background audio will almost certainly give the game away; the answer is a simple crossfade between the adjacent audio tracks, as little as six frames could be more than sufficient to soften the effect and create a truly seamless edit.

Crossfades are applied in different ways

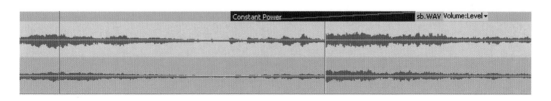

Application of a dedicated crossfade transition effect in Adobe Premiere Pro.

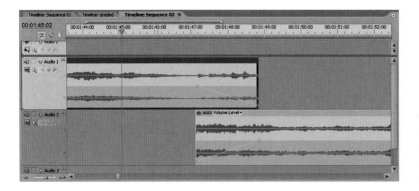

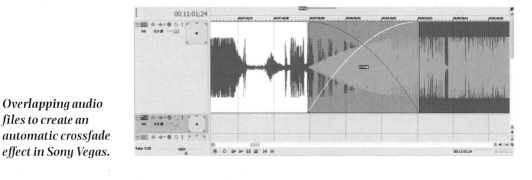

Another approach – rubber banding audio levels of files on adjacent timeline tracks.

Overlapping audio files to create an automatic crossfade effect in Sony Vegas.

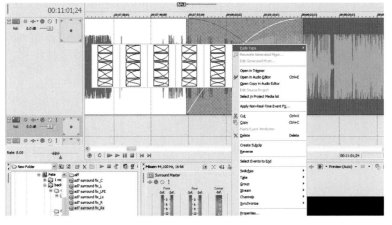

Sony Vegas also offers a selection of acceleration curves for audio transitions and fades.

depending on the editing program you use. In Adobe's Premiere Pro and Premiere Elements, for example, audio transitions are found in a dedicated effects palette and dropped direct between adjacent sound clips. The effect's duration can then be altered by dragging on the timeline or setting a numerical value in a control panel. In other programs, such as Sony's Vegas or Magix's Movie Edit Pro, audio transitions are set by overlapping two clips on the same timeline track. In the case of Vegas, a rubber band display is created to represent the fades in and out and the shape of these bands can be altered to create a sense of acceleration from one to the other. With other editors it is often necessary to create crossfades manually, by setting audio levels individually for adjacent clips, using the program's mixer panel or rubber bands.

AUDIO FILTERS AND SWEETENING

Filters alter the tone and the sound of audio. They can be used to clean up or improve sound quality, removing hiss, enhancing bass or treble frequencies or compressing the dynamic range to avoid extensive changes in volume. They can also be used for more wild and conspicuous effects, such as altering speed or pitch, introducing echo and reverberation or degrading quality to emulate the sound of devices such as telephone receivers.

Clean-up filters present in entry-level editing programs include the likes of the

> ### Elaborate Audio Transitions
>
> While audio doesn't lend itself to a great array of transition types in the same way as video, there are ways to pack more of a punch if subtlety isn't a priority. Consider simple, abstract sound effects such as a swoosh, or more recognizable sounds such as that made by a passing car, if we are cutting to a roadside location. Simple instrumental notes can add drama to a transition (or 'cheese', if you do it badly). Read ahead to find out more about spot effects and think about applying them to scene changes.

removal of wind noise and hiss. The quality of the job they do can vary greatly depending on the nature of the recording, and there is no one-click solution for enhancing every dodgy recording you're likely to make. Hiss, buzz and hum removal, for example, work by

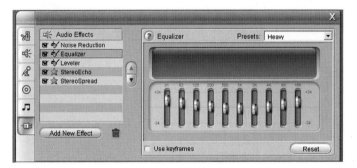

ABOVE: Audio filters stacked up for a cumulative effect in Pinnacle Studio.

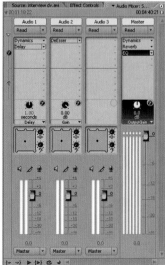

RIGHT: More advanced editors, such as Premiere Pro, allow sound filters to be applied to entire tracks or even the finished mixed output.

suppressing certain sound frequencies commonly associated with wind, tape hiss or electrical interference. Some of these frequencies may also be present in noises you want to keep, however, and excessive use of these filters can leave you with a flat, tinny or just plain unnatural sounding audio track. In more advanced programs, editors are expected to understand how to use the central audio effects tools, such as equalizers and compressors. Each audio filter will often come with a selection of presets to help you to match the job you're trying to do, but they will ultimately need some fine-tuning. And, even then, there is no guarantee that you can save a poorly recorded soundtrack.

Equalizers

Equalizers provide control over a specific frequency range. In a graphic equalizer, frequencies are divided into three, ten, twenty or more bands, each one controlled with a slider. Graphic equalizers are common on home stereo systems and take the same visual appearance in video editing programs (we have already encountered the idea of virtual sliders when discussing mixer panels).

Parametric equalizers are more versatile but much more complicated. While a graphic equalizer is clearly laid out to provide access to frequencies running from low to high, a parametric equalizer asks users to specify a frequency range and the size of that sample and to adjust it as required. They permit a lot of delicate control but are tricky to understand and manipulate. What is more, equalization, like sound mixing in general, is a precarious business: by enhancing one sound in a recording, you risk unbalancing countless others.

Noise Gate Filter

Also available in prosumer software and worthy of investigation is the noise gate filter. Designed to remove background noise from recordings, the effect removes all sound that falls below a certain volume, rendering that portion of the soundtrack mute. If the problem is hiss or electrical interference it will still be present on sections of recording that include speech, music or other stuff that you want to keep, but it will be lost from quiet sections. Used in anger, a noise gate can lead to very unnatural results as sound comes and goes like

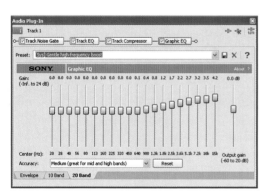

A graphic equalizer in Sony Vegas.

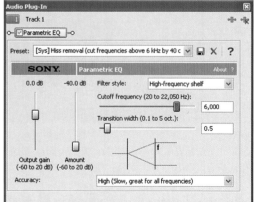

A parametric equalizer.

a speaker with a loose cable. The effect is easy to mask, however, if you have been clever enough to record wild tracks (covered in my previous book and mentioned briefly below).

Normalization Filters
These are useful if recordings are particularly quiet. Normalized audio amplifies sound in such a way that the loudest peaks are pushed as high as possible, while the quietest moments remain quiet. Normalizing audio is a good way to boost volume without also amplifying background noise or compressing the dynamic range. On the negative side, a normalization filter will place the loudest noise samples at the highest level (possibly 0dB), make the quietest totally silent, and assign values to all other samples in relation. The result can be extreme changes of volume between differ-ent parts of the recording that appeared to have little variation before the effect was added.

Limiting
Another tool for levelling out the audio levels on a soundtrack is limiting. Here, a volume limit is specified – possibly 0dB or –6dB. All audio samples will be made to peak at that level. A hard limiter takes a ruthless approach to this process, creating a neat flat line of sound peaks at the required level. Applying a limiter is a working solution in bringing uniformity to sound levels during a recording, while avoiding the expanded dynamic range that can occur during normalization. On the downside, heavily limited audio can sound flat, dull and lifeless.

Compression
This is another useful tool that can save or kill audio quality. A compressor squeezes the dynamic range of a sound file so that all samples fall into a specified volume range. There are many different ways to compress a sound file, depending on the nature of the

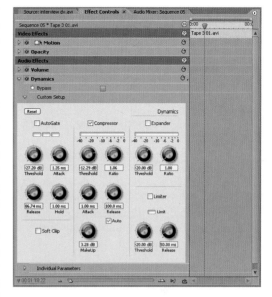

Compression tools in Adobe Premiere Pro.

source material and the effect you are trying to achieve, but most good editing programs at the prosumer and professional level provide template settings to suit common scenarios. Opinion is divided on when (and even if) compression should be applied to a sound mix. Many save compression to the very end of a sound mix and apply it to the output channel in order to create a tighter sense of integration between the component parts. Others apply compression to source files prior to a mix in order to make the files clean and uniform before editing. Either way, it is important to note that too much compression will yield an unnatural 'breathing' effect as sound levels rise and fall to compensate for natural shifts in intensity. When a subject stops speaking, for example, the background audio will get louder.

Other Techniques
Taking a more stylistic approach, some editors provide filters such as pitch shifting (raising or

Garbage In...

Sound effects added in video editing and audio editing programs are easily removed if you don't like them. Background noise, echo and digital clipping present on the original source recording are much more difficult to remove, if not impossible. Forget what you've seen in those high-tech police thrillers, where a technician strips layers of background noise from an emergency services call to isolate the sound of the caller's budgie and determine its accent. Here, in the real world, sound recordings are not nearly so malleable. Think of it like trying to separate the ingredients of a cake after it has been baked. If you want full control of your sound quality, record it cleanly. Think about what you can add later, not what you will try to remove.

lowering audio pitch without altering its speed) and reverb (adding echo to a recording to simulate the effect of a large room or hall). Going even further and taking a giant step away from the world of subtlety, strange effects such as robotic voice filters give you altogether daft results that you may or may not be able to put in context with your movie.

USING WILD TRACKS

If you have read *Digital Camcorder* you will be familiar with the idea of recording wild tracks. In brief, a wild track is a rather nondescript recording of ambient background noise. Each setting and location has its own background chatter, and selecting the right type of noise can have an immense impact on the atmosphere of your location. Imagine a restaurant scene: a conversation scene cut with a man and woman at a dimly-lit dinner table will have a quite different feel if cut with a sedate background track of quiet, indistinct chatter and soft piano music than it would have if the background audio was that of a drunken works' night out with bawdy songs and breaking glass.

Applying a wild track to your video not only adds atmosphere and sets the scene in context, but it fools the viewer into believing the real-time nature of the events being portrayed, regardless of how many edits are involved nor how long it took to shoot the scene. Nothing hides video edits quite as effectively as subtle and continuous sound. Don't try to draw attention to wild tracks, they work sub-liminally and should really not be noticed on a conscious level. If your wild track recordings are made with good high audio levels, consider dropping the volume considerably. Your goal here is to add depth to a setting rather than overpower it. The wild track should sit comfortably in the background and sound as if it could have been recorded with the picture rather than added later. A carefully recorded wild track should have no discernible speech nor recognizable words. Similarly, there should be no identifiable sounds, such as chiming bells, slamming doors or shrieks of terror. In most cases, a wild track will last for only one or two minutes, so it may be necessary to loop it over longer scenes. This is where identifiable noises can kill the illusion if they are allowed to repeat more than once. When looping a wild track ensure that all occurrences remain at the same volume and add a short audio crossfade to the join between the loops in order to avoid sudden and audible cuts.

ADDING SPOT EFFECTS AND FOLEY

Background ambience isn't the only thing you will have to replace or recreate in a video soundtrack mix. Spot effects are also

Apple's Soundtrack Pro can be used for general audio mixing or for the creation of wild tracks and foley audio.

important – that is, sounds created by physical things or actions, such as clapping hands, teacups being put down on saucers or footsteps on gravel. If you have been recording sound to get the clearest possible speech it is more than likely that incidental sound effects have been lost along the way and will need to be replaced. But, in a way, that works to your advantage too since the replacing of spot effects allows much more freedom with regard to mixing audio levels and ensuring that sounds add detail and context to a scene without overpowering the main dialogue.

Foley is a term normally associated with the recreation of footstep noises, but it can also be applied to other types of spot effect, including, but not limited to, opening and closing doors or breaking windows. The art of foley involves

Spot effects can also be found in beginner's editing programs, such as Pinnacle Studio.

creating and recording sounds that sound like an object or action but are not actually made by it. It is a very common procedure in the film and TV industry, but its roots stretch back to radio drama. A foley studio will often have a selection of surfaces and types of shoe with which footstep sounds can be made. It will also have specialized props designed to make particular noises. A small wooden box will serve as a door and a jam jar full of metal tacks might be used for a breaking window. Rumour has it that the ominous sound of a Martian spacecraft opening in Byron Haskin's 1953 adaptation of *War of the Worlds* was actually made by unscrewing the lid from a coffee jar. But, however sound effects are made, the recordings must be clean so as not to introduce additional background chatter into the sound mix. Audio levels must also be carefully considered when foley effects are added to a mix. In most cases the sound should be loud enough to be audible, but not so high that it grabs the viewer's attention. On the other hand, pushing up the volume of a sound effect can give it a very imposing feel, as with Haskin's coffee jar lid.

Commercial CDs are available providing sound and foley effects for specific settings and themes.

As an alternative to recording foley effects with a microphone and selection of knick-knacks, there is also the option of buying sound effect CDs. They are often collected into themed compilations, with horror, war or action themes and can vary greatly in quality. Collections used by professional production companies tend to be rather expensive, but relatively cheaper than hiring a foley studio to have sounds custom-made. If that doesn't appeal, a final option is to make use of sampling software and a MIDI keyboard. Many music-sampling programs support sound effect samples as well as virtual instruments, allowing editors to 'play' footsteps and other sound effects into their soundtrack by using a musical keyboard.

A NOTE ABOUT MUSIC

Be careful when adding music to your videos; adding pre-recorded commercial music to your movies without written permission from the publisher can be an infringement of copyright and lead to all sorts of legal nastiness. It is true that any videos you make for your own personal use are difficult – if not impossible – to police, but for any jobbing work, however, where video is being made as part of a chargeable service, extra care must be taken to stay on the right side of the law. If you can't afford the cost of mainstream commercial music (and don't have the skills or resources to write and record your own) two options are open to you. First, you could look at the many copyright-free CDs that are available for video producers. These tend to be pricey but are much cheaper than paying royalty fees to record labels and music publishers. On the downside, library music CDs cater for a low common denominator and as such, their offerings are often bland and lifeless, offering little emotional support for your visuals. They are fine for links in magazine-style videos,

SmartSound QuickTracks integrates itself into the main video editing interface and provides a choice of composition, arrangement and style, ultimately creating a musical piece that snugly fits the desired duration.

sports events and corporate presentations but bring little of worth to drama or weddings. Try to visit the websites of stock music publishers and listen to sample tracks from each CD before parting with your money.

The second route to take is music generation software. There are three popular contenders in the DV mainstream, each with its own way of working. One of the most commonly found music generators is Smart-Sound's QuickTracks, which is provided as a plug-in for many video editing programs from the entry level up. QuickTracks allows users to specify the duration of audio they need and to choose from a selection of musical styles, compositions and arrangements, at which point a piece of music is created to neatly fit the required duration. The 'free' version that comes with many editing programs has a reasonable range of songs and arrangements, but it won't be all things to all people. Smart-Sound is, however, hoping that you will log on to its website and download more content to better suit your needs – and, in truth, that is a great way to acquire stock music.

The second program worth mentioning is Sony's Cinescore. It is designed to work alongside the company's editing program Vegas, but works just as well on a stand-alone basis. In this case, music is cued up on a

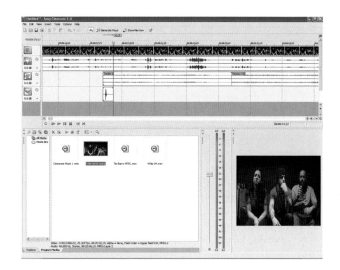

Sony's Cinescore provides a more hands-on, timeline-based environment in which users can work. Composition and arrangement options are still good and users can opt to manipulate the volume or the intensity of the music at specific times to fit the drama unfolding on-screen.

Apple's Garage Band is a superb program for the money and can be used as a fully functional soundtrack scoring application, being compatible with USB musical keyboards as well as ready-prepared loops.

timeline along with the video track itself. As with SmartSound, users have a choice of song, style and arrangement, and Cinescore assembles a musical track from a selection of component parts. Their intensity and volume can be altered direct on the soundtrack, alongside marked cues, all of which help to synchronize the tone of the music to the events on screen. As with QuickTracks,

Nesting Timelines

Good editing programs at the prosumer level allow timelines to be nested one inside another, meaning that a movie you have just edited can be added to a new sequence and treated as a self-contained video clip without the need to render it to the hard drive as a self-contained file. The feature is also useful with sound – audio can be mixed in one timeline, then added to a new (or duplicated) sequence where it is treated as a self-contained sound file. The approach is useful when you are dealing with a lot of separate sound elements.

more music samples can be bought from the company's website.

On the Mac platform Apple has catered well for its customers with Garage Band, a music sampling program that now supports a video track and allows music to be added in sync to movies. The program supports external keyboards and even real instruments connected via analogue ports. It also comes with a good selection of ready-made percussion and instrumental samples that can be assembled and arranged to suit your needs. And, of course, more instruments and samples can be bought from Apple in the form of 'Jam Packs'. For such an inexpensive program, however, Garage Band is phenomenal and can be as simple or as hands-on and complex as you want it to be.

EXTERNAL SOUND EDITORS

Often, a video editing program will not provide all the audio tools you need. This is most likely to be the case when working with a budget, entry-level editor. Luckily, there is a good range of dedicated audio editors available, at prices ranging from a few to a few thousand pounds.

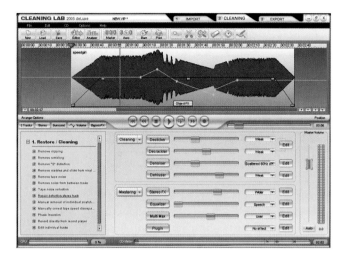

Magix Audio Cleaning Lab Pro offers excellent sound editing tools for very little money.

Some CD-burning applications come with simple audio editing tools, and programs such as Magix's Audio Cleaning Lab provide quite sophisticated equalization and compression tools at a very keen price, as well as the ability to destructively import, filter and save a video file's audio track, if need be. Moving up, you will find applications, such as Adobe Audition, which allow multi-track sound editing with surround-sound panning capabilities. The range of programs is vast and audio software can cost thousands of pounds if you are keen enough to invest in the top-end of the studio-recording market. External sound editors may feel like a distraction from the movie-making process, but it is useful to have something on hand if the main editor's toolkit comes up short.

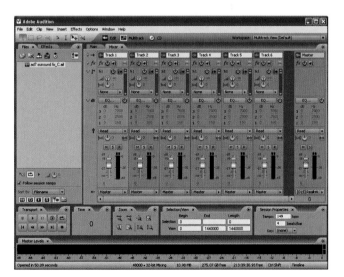

Adobe's Audition is a more advanced audio editor, intended to complement a suite of applications, such as Premiere Pro, After Effects and Encore DVD.

7 SPECIAL EFFECTS

Considering the emphasis that video editing software developers place on video effects, it may seem odd that I choose to cover them so late here. There are two straightforward reasons for this: first, the structure of the text is intended to follow video post-production in the order in which it would be tackled in most real-life situations. And, secondly, video effects are not as important as advertisements and promotional websites would have you believe. Video effects can provide you with a distraction or help to enhance the mood and tone of a movie, but they don't bring an awful lot to structure, pacing or performance,

---

When You Can't Cut

Another maxim you will often hear among video makers is 'cut, cut, cut, and, when you can't cut, dissolve'. The message here is plain – don't use transition effects unless you really have to, and, even then, restrict yourself to the most subtle. The more elaborate a transition effect, the more you show your own hand and the more the audience is distracted from the on-screen content. Always remember that fancy effects are not key ingredients for a killer movie.

---

elements which should be your primary concern in storytelling.

Video effects fall into a number of different categories. The most commonly used are transitions (which ease the change from one shot to the next) and filters (which affect the appearance of footage). In addition, video editing programs will also provide some kind of titling tools and may even go to the lengths of allowing keying and compositing (methods of combining multiple video and image files within a frame). Some video effects, such as colour correction and grading tools, have a very subtle effect and serve to enhance the emotive quality of your work, while others are so wild and wacky that they serve only to draw attention to themselves. They can all be useful if used in the right context, but be wary of letting them dominate. Think of effects like spices in cooking – a little can make food interesting, but add too much and all you will taste is the spice itself.

TRANSITIONS

Just like the audio crossfades discussed in the previous chapter, video transitions provide a gradual change from one clip to the next. The most commonly used transition in film and TV production is the cross dissolve, which has the outgoing clip fade away as the incoming one fades in. It is undoubtedly the most subtle

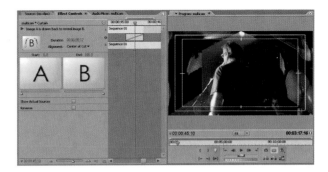

Transition effects on display in Adobe Premiere Pro.

transition you are likely to find in the vast selection provided by most video editing programs. Less commonly used, but still regularly seen is the wipe, in which the outgoing clip is wiped away with the in-coming one. The direction and the speed of the wiping action can be customized, as can the shape of the boundary between the two images. This border can also be softened or given a solid, coloured edge. Wipe transitions were used as a signature effect in the *Star Wars* movies, but are often thought of as being too gimmicky for straight drama.

But that is only the tip of the iceberg though, and a quick inspection of any video editing program at any level will reveal a massive treasure trove of transition effects with extraordinary properties such as page

A selection of stills from a simple cross dissolve transition.

A no-frills wipe transition.

A page peel – one of the more extreme transitions that can be found in most video editing programs.

peels, spins, flips, explodes and zooms. The nature of these effects has been pushed to extremes as software programmers learned how to tap into powerful graphics cards intended for rendering 3D games. Such effects are technically sophisticated and great fun to play with, but you will be hard pushed to find a serious context in which to use them in most of your video projects. On no account should you use them 'just because they're there'.

In most editing programs transitions are dropped between adjacent clips on a single video track of the timeline or between adjacent frames of a storyboard. In a timeline-based editor, the duration of the transition can often be altered by dragging its edges left or right, and there will also be the means to set the duration numerically if required. In a small number of editors the main editing track is divided into two – an A roll and a B-roll. In this model, clips are placed on different rolls in

such a way that they overlap. The transition effect is then placed in the overlapping portion. Such editors are now becoming quite scarce, but the AB-roll model is still evident in the control panels of programs such as Adobe Premiere Pro, used for fine-tuning transition effects.

While you are unlikely to encounter the AB-roll model for transitions, it is useful to be able to visualize them in that way. In particular, think about the 'overlapping' media – if there isn't footage to spare on either side of a cut you might not be able to apply a transition at all. And you also need to be sure that the effect doesn't run into a call of 'cut', in which the camcorder might wobble as it is put down, actors look direct at the camera or drop out of character. Remember too that in most editors transitions can be placed at the beginning, middle or end of a cut.

The transition's control panel will also

Transition effects in Final Cut Pro, dropped between clips on the same timeline track.

An AB roll structure still exists in some editors' effects controls (as seen here in Final Cut Pro), but not on the timeline itself.

provide settings for any crazy effects you've chosen, such as the shape, direction and speed of wipes, or their border colour and softness. The more elaborate an effect is, the more settings there will be to tweak.

VIDEO FILTERS

Video filters affect the general nature of a video clip. They can be used for subtle effect in the form of colour correction or for more stylistic purposes by using a huge selection of warps and distortions that come with practically every video editor on the market today. Colour correction is undoubtedly the most useful form of video filter you will encounter – not just for correcting mistakes but for enhancing the mood of a piece. And, because of its importance, we shall be covering colour correction tools in isolation shortly.

No matter which video-editing program you are using you will almost certainly be in command of a lot of wild-looking video filters. To help to make to sense of them all, many programs organize them into categories depending on their purpose, examples might be image control, stylize or distort. In addition, you might be presented with thumbnail images, offering a basic representation of what an effect looks like. The selection of filters can differ from one program to the next. They will all have some form of image correction tools

that we shall cover later, and you are also likely to get a mosaic filter in which the picture is composed of large coloured blocks. In addition, common effects include blur and sharpen filters and old film-look effects (to simulate the colour, grain and damage associated with old cine reels). You are also likely to find weird colour distortion effects such as a solarize filter, as well as general image distortion effects used to simulate lens

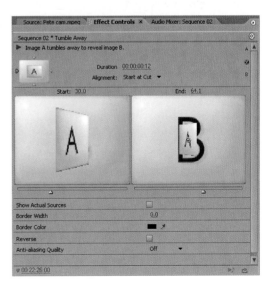

Extensive control options are often available for the wildest of transition effects.

83

Even beginners' programs such as Ulead VideoStudio offer a huge selection of video filters.

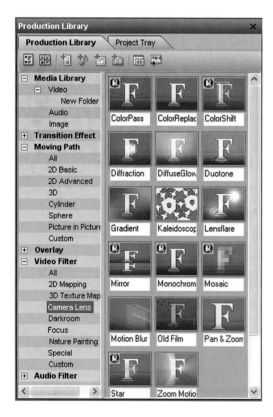

LEFT: An even richer assortment is available in more advanced programs such as Ulead's prosumer offering MediaStudio Pro.

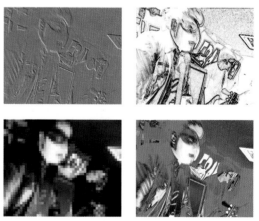

A selection of some of the most extreme (and arguably least useful) video filters you are likely to encounter.

84

distortion or the look of shooting through running water. More advanced editors may even provide half-tone effects that resemble the dot-based structure of newsprint. There are more effect options in the video editing market than I would have the space (or patience) to list here. In some editors – usually only those at the prosumer level – effects can be keyframed so that their individual properties can be changed or intensified over time. They are fun to experiment with but, as always, use them with respect. Always make sure that any effect you add to a video clip has a purpose and is being used to enhance the point of the video rather than just to show off your favourite editing toys.

Typical brightness and contrast controls.

COLOUR CORRECTION

Colour correction and image controls are, without doubt, the most important video filters available to you. Even if your video is impeccably well shot, with technically perfect white balance, you might still want to make changes to the colour quality, contrast and brightness in order to evoke a sense of emotion and atmosphere. The process is known in the movie industry as 'grading' and is an essential part of the post-production process in all mainstream dramas for cinema and TV. Even at the entry level, there is a lot you can do to change the tone and feel of your videos, and the colour correction tools in more advanced programs can be quite intimidating.

At the most basic level a video editing program will give you the option of changing a clip's brightness and contrast levels. These are controls that anyone with a television set should be familiar with: brightness controls dictate the total amount of white in an image, while contrast determines the distance between absolute white and absolute black, generally making the image more delicate as the contrast is reduced or more robust as it is

increased. As an alternative to contrast controls, more advanced editors, such as Adobe Premiere, provide levels controls, which allow a much more specific placement of white and black within the image and can often be applied to individual colour channels.

In addition, you are likely to find hue controls which move the video's colours around the colour wheel. When used subtly, this has the effect of making the general tone warmer or cooler by weighting the colour palette more towards the red end or the blue end of the spectrum. At its extreme, however, you can end with completely incorrect and artificial-looking colours.

Along with hue controls, you will also be given a saturation slider, which affects the total amount of colour within the video. Again, this is similar to the colour control on most TV sets, serving to make the picture more vibrant (often at the expense of detail) or more subdued – with the extreme point being a completely black and white image.

A slightly more advanced approach to colour correction comes when a program offers the independent control of colour saturation in red, green and blue channels. This is where more creative decisions can be made and where you can start to match the tone and quality of footage that comes from different sources or different camcorders. White balance can also be corrected by using RGB colour controls, although some of the best editors also provide a specific white balance correction tool, which adjusts all the colours in an image, taking reference from part of the frame that the user identifies as 'white'. And at the most advanced end of the prosumer market, we have three-way colour correctors, which provide comprehensive controls over individual colour channels in the picture's highlights, mid tones and shadows. These complex tools also come with histograms and waveform displays, offering a purely objective view of the video's brightness and colour content. They can be difficult to

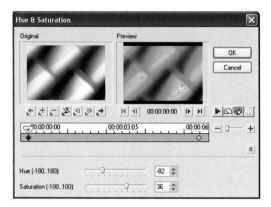

Hue and saturation controls for controlling colour warmth and intensity. They often also come with a lightness control for altering the total amount of white in the picture.

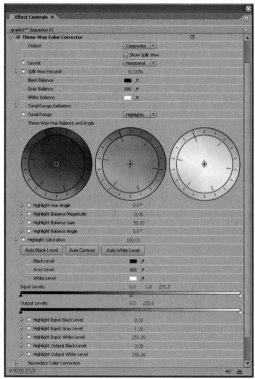

Comprehensive colour correction tools as found in Adobe Premiere Pro. A similar toolkit can also be seen in Final Cut Pro and Vegas.

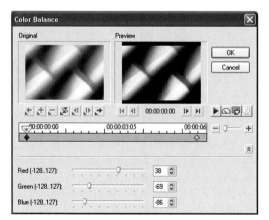

Independent control over a video clip's red, green and blue channels.

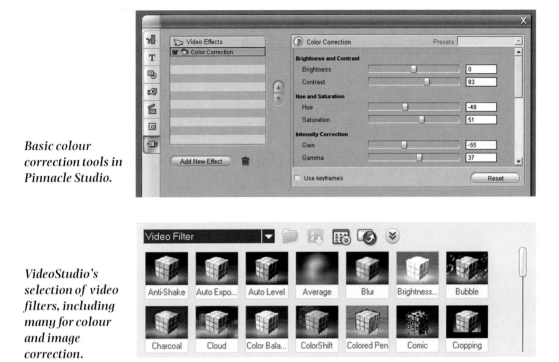

Basic colour correction tools in Pinnacle Studio.

VideoStudio's selection of video filters, including many for colour and image correction.

read and understand, but they are a huge help if you ever need to match the colour tone of two very different-looking clips.

Entry-level Image Correction – Ulead VideoStudio

When a clip is selected on VideoStudio's timeline while in edit mode, the video panel offers a button marked 'colour correction'. From here, you are just one click away from a simple control panel with sliders for adjusting hue, saturation, brightness, contrast and gamma.

Results are updated in real time in the program's video viewer, to give an immediate impression of how the finished results will look. More image control options are found among the program's video filters – but you will not find them in VideoStudio's Effects interface. Stay in edit mode and click the

attribute channel on the main control panel, that done, the gallery panel will display a selection of video filters (represented as animated thumbnails) while the attribute panel itself offers up an empty pane in which effects are stacked and their order changed if need be.

Auto levels and auto exposure act as one-click solutions to enhance video. But they might not give quite the results you want and their parameters can't be customized. More control is given over a brightness filter, which offers sliders to adjust the overall brightness, gamma and contrast. Similarly, the colour balance filter gives independent control over the saturation of the red, green and blue channels. There is an enhance lighting option that simulates fill-in flash by pulling detail out of shadows and can also be used to accentuate shadows if required. And, just in case you

87

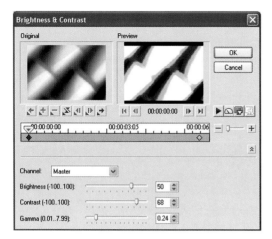

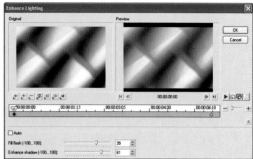

ABOVE: **VideoStudio's enhance lighting options.**

LEFT: **Brightness filter controls.**

didn't see the original colour correction tools, the filters gallery also includes dedicated hue and saturation controls.

Intermediate-level Colour Correction – Premiere Elements

One of the easiest methods of enhancing dull or uninteresting footage in Premiere Pro is to use the auto levels controls. Here, the software will automatically determine absolute white and black values within the image and adjust its tone and contrast accordingly. Unless you are going out of your way to create a distinctly overexposed or dark-and-gloomy look, this simple, one-step filter will yield good results almost immediately.

Fine tuning controls include temporal smoothing – in which the program is made to

Auto levels options in Premiere Elements.

analyse frames on either side of the one it is adjusting to ensure that there are no wild fluctuations as the clip plays out. There are also a white clip and a black clip setting which determine the amount of detail that will be left in extreme shadows and highlights. And, if the final result seems too extreme, a blend option is on hand to blend it with the original footage.

More hands-on corrective tools include brightness and contrast, and, while these may seem the most obvious options for brightening up footage, they can ultimately weaken the effect overall. To enhance detail in the mid tones without compromising shadows and highlights, try using gamma correction instead. Alternatively, the shadows/highlights filter can be used to adjust the highlights and shadows without significantly affecting the mid tones. Hue and saturation controls are found in the settings for a filter called image

RIGHT: **Brightness and contrast controls.**

BELOW: **Gamma adjustment.**

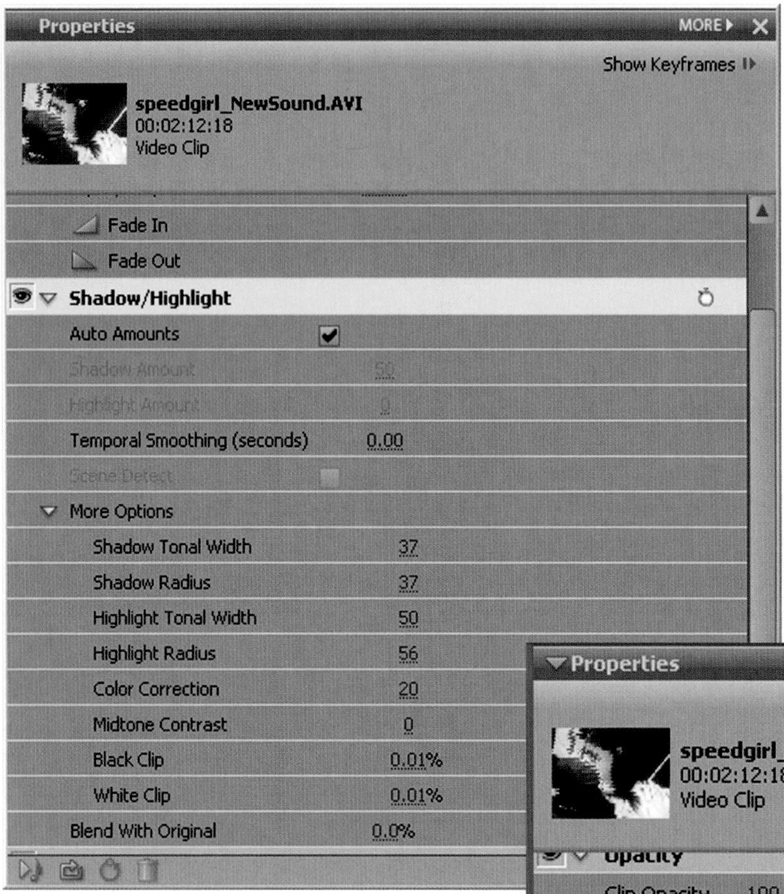

Properties MORE ▶ ✕

Show Keyframes ◄▶

speedgirl_NewSound.AVI
00:02:12:18
Video Clip

◢ Fade In

◣ Fade Out

👁 ▽ **Shadow/Highlight** ⟳

Auto Amounts ✔

Shadow Amount 50

Highlight Amount 3

Temporal Smoothing (seconds) 0.00

Scene Detect

▽ More Options

Shadow Tonal Width 37

Shadow Radius 37

Highlight Tonal Width 50

Highlight Radius 56

Color Correction 20

Midtone Contrast 0

Black Clip 0.01%

White Clip 0.01%

Blend With Original 0.0%

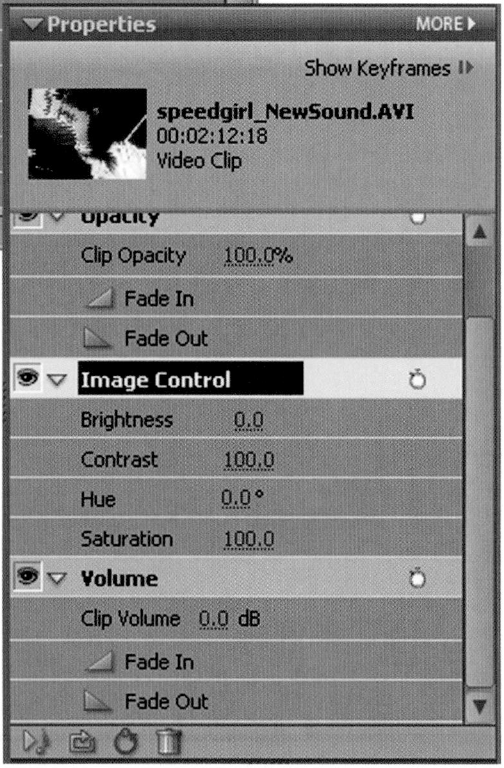

Properties MORE ▶

Show Keyframes ◄▶

speedgirl_NewSound.AVI
00:02:12:18
Video Clip

👁 ▽ **Opacity**

Clip Opacity 100.0%

◢ Fade In

◣ Fade Out

👁 ▽ **Image Control** ⟳

Brightness 0.0

Contrast 100.0

Hue 0.0°

Saturation 100.0

👁 ▽ **Volume** ⟳

Clip Volume 0.0 dB

◢ Fade In

◣ Fade Out

ABOVE: Shadows/highlights filter.

RIGHT: Hue and saturation controls.

control. A very slight shift of colour hue can give an image a subtly warmer or cooler tone. And in shots where detail is scarce, a slight drop in colour saturation can help greatly.

Advanced Colour Correction – Final Cut Pro

Final Cut Pro, like most video editors at the prosumer level, has extensive colour correction tools that can dig you out of substantial holes, help to evoke a sense of atmosphere in your movie and also confuse you greatly if you don't know where to begin. There is no substitute for reading the manual, but here is a basic introduction to the core toolkit for three-way colour correction.

Once the video has been edited and all clips are present on the timeline, locate any one that needs adjustment and select it with a single mouse click. If a whole sequence needs to be adjusted load it into a new timeline, where it will be treated as a single video clip. Colour correction tools can be found under two categories of Final Cut's Effects browser: colour correction and image control. Image control and colour correction filters are dragged over to the clips and nested sequences on the timeline and are applied immediately. Once done, click the viewer's filters tab to gain access to the effect controls. Image control filters are designed to affect specific properties such as brightness and contrast and have been addressed earlier in this chapter. Final Cut Pro's colour correction tools are more complex and will be discussed here.

Final Cut Pro's Three-Way Colour Corrector

Final Cut's effects browser.

can be viewed in a visual interface, which presents users with three colour wheels, or in numeric mode, with quantitative values for each property. Either approach allows colour adjustments to be made independently to shadows, highlights and mid tones. Each colour wheel in the Three-Way Colour Corrector's visual interface affects hue and saturation. Moving a small, round balance control indicator in each wheel makes changes; altering the angle of the indicator affects hue, while its distance from the white centre controls saturation, each wheel allows changes to be made in shadows, highlights or mid tones, and holding down shift while dragging moves the indicator in a straight line from centre to edge. The holding command makes adjustments quicker.

The Three-Way Colour Corrector also provides a level slider for whites, mid tones and blacks, permitting greater control over image contrast. Dragging the whites slider to the right will brighten highlights, helping an underexposed image, while pulling the blacks slider to the left drops more detail into shadow. There are also three auto levels controls at hand: auto white finds the brightest highlight in the clip and assigns to it a white level of 100; auto black finds the darkest shadows and gives them a white value of 0; auto contrast sets both controls at once. Another slider in the colour corrector panel is labelled sat, short for saturation. The sat slider affects the overall colour saturation for a clip across highlights, shadows and mid tones. If you are tempted to boost saturation, enable range checking to

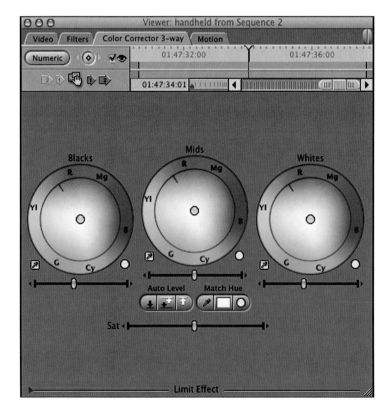

The Three-Way Colour Corrector.

Colour matching is a feature exclusive to advanced editors in the prosumer market and above.

ensure that they are not pushed too high – see the box that follows for more on range checking. Colour balance can be adjusted automatically by using a small eyedropper icon at the bottom left of each colour wheel – called the auto-balance colour function. Once selected, users can select colours by clicking them on the actual video image. Use the white wheel's eyedropper to select an item in the video that is supposed to be pure white. Similarly, the black wheel's is used to identify a black element. Colours will be automatically adjusted to compensate.

One feature that Final Cut Pro offers that you are unlikely to find in entry-level editors is a dedicated interface for matching colour properties between two different clips. Imagine that footage has been shot with different types of camcorder or that some of the footage for a scene has been incorrectly white balanced. Rather than follow a long path of trial and error, here's the program's approach for accurate matching: in Final Cut's Windows menu select Arrange>Multiple Edits. You will need access to a reference clip (with correct colours) and a target clip which is to be adjusted. The reference clip is selected by using drop-down menus at the bottom of frame viewer 2. In the colour correction or three-way

colour correction editor, click the eyedropper icon under the words 'Match Hue'. In the reference frame, click on a colour that you want to match. Notice that the selected colour appears in the match colour palette, and one of the colour wheels' eyedroppers becomes highlighted in green. Click the highlighted

Copy and Paste Effect Settings

It can take some time to adjust the colour and tone of a video clip correctly and the task of applying the same settings to multiple clips within a movie can seem a daunting one. If several clips need to receive the same adjustments, however, these can be applied with a simple copy and paste action. Highlight the required attributes in the filters panel, use a standard copy command, then select the target clip and choose paste. Alternatively, click the drag filter icon at the top of the editor and drag it to another clip on the timeline. You can also select an adjacent clip and use copy from the 1st clip back commands to apply the same settings automatically.

Range Checking

Final Cut Pro's range check features serve as a constant reference to a video's safe luma (brightness) or chroma (colour) limits. Enabling range checking is done from the View>Range Check menu and has the options of displaying excess luma, excess chroma or both. Areas with signal peaking show red and green zebra stripe patterns – red for luma and green for chroma, if both are being monitored, otherwise green stripes represent values that are approaching illegal levels, while red indicates areas that have exceeded them. A green tick in the top third of the canvas indicates that footage is still legal, while a black exclamation mark in a red triangle indicates that problems need to be addressed.

eyedropper. In the clip you want to adjust find an element which is supposed to have the same colour and tone as the item you clicked in the reference clip. Clicking on it will identify it as having the same shade and colour, adjusting the clip's colour properties accordingly.

CREATIVE COLOUR CONTROL – SONY VEGAS

Sony Vegas takes a rather unorthodox approach to video editing, but its capabilities in sound editing are unparalleled in the prosumer video market. It also shines brightly in its effects tools too, as we shall see in this short exercise in colour design. The following pages will briefly introduce Vegas's colour correction tools, highlighting similarities to those of Final Cut Pro above and also explain the use of its secondary colour editor in isolating individual colours for extreme effect.

In Vegas effects can be applied to an individual clip or to an entire video track. To work on a clip-by-clip basis, right click on the required clip and select Video Event FX from the context-sensitive menu that appears. To apply an effect to an entire track, click the button named Track FX in the track head. In either case, a window will appear offering a vast selection of effects. First, take a look at the colour corrector; as with Final Cut Pro, you'll notice that it has three colour wheels – one controlling shadows, one for mid tones and one for highlights. They are operated in exactly the same way too, by positioning a small circle at an angle to adjust hue or moving it from the centre to the edge to increase saturation. Each adjustment that can be made visually with the colour wheel can also be done quantitatively by using numerical controls. Eyedropper buttons in Vegas's colour corrector work slightly differently from those in Final Cut. With Vegas there are two buttons for each wheel, one with a plus and one with a minus. The 'plus' eyedropper, on the right of a wheel, allows users to choose an adjustment colour from the source footage. Click the eyedropper and then move the cursor over to the video display and select a colour from the current frame that you want to boost in that particular wheel. The eyedropper with the minus symbol helps to remove chosen colours from the image, this is done by boosting the complementary colour, located at a 180-degree angle on the colour wheel. As before, click the eyedropper button and then choose a colour in the video frame that you want to diminish. A set of sliders that affect image properties across shadows, mid tones and highlights together include general saturation controls, gamma adjustment, gain control (which affects brightness) and 'offset', which serves as a highly sensitive contrast control.

Vegas's secondary colour corrector enables adjustment of individual colours within an

Sony Vegas's main Colour Corrector.

The secondary colour corrector is ideal for more stylized colour-adjustment effects.

A visible mask indicates how accurately the area to be affected has been isolated.

95

image, while leaving the others unaffected. It features only one colour wheel and a lot of sliders and numerical controls. With the secondary colour corrector selected and applied to a clip or track, select the effect range eyedropper and choose a colour in the video viewer. If the colour has a graduated tone, hold down the mouse button and drag a rectangle to select an area with which to choose a range rather than a single pixel.

Clicking an option labelled 'show mask' will display a black and white image showing selected and unaffected areas of the frame. This is a great help in illustrating how well you have isolated your subject. With most video footage it is unlikely that these first few steps will yield a clean and precise mask, so additional adjustments will be necessary. Check the boxes marked limit luminance, limit saturation and limit hue. These controls allow you to widen the tolerance of pixels and also apply a degree of softness to the mask edge for

each parameter. Once you are happy that all of the target area has been isolated and that there is no significant overspill of the mask into other parts of the frame, use the secondary colour editor's main controls (at the top of the panel) to adjust colour in the selected area. There are controls for hue, saturation, gamma, gain and offset, as well as an alpha slider to make the selected area transparent for compositing. The mask can also be inverted so as to make changes to everything but the selected colour. In this case, we have decided to isolate one brightly coloured item within the image and to drop the rest of the frame to black and white. As with all effects filters in Vegas, colour correction and secondary colour correction tools can be keyframed, allowing a change in colour properties over time. The very bottom of the effects editor features a scrubbing bar into which keyframes can be placed.

About Video Scopes

Video scopes provide an objective representation of on-screen content, which is useful when you consider the different ways in which computer monitors and TV sets can be configured. In particular, pay attention to the vectorscope, which is laid out in the same manner as the colour corrector's wheels.

Video scopes may seem abstract and difficult to read, but they provide a good objective representation of the video's colour and tonal values.

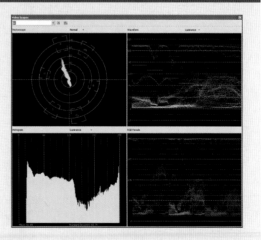

COLOUR GRADING – MAGIC BULLET FILTERS

If there is such a thing as a Holy Grail in the world of digital video editing, it is the goal of making movies look as if they were shot on film. It's not that video looks bad, it is just that the big budget movies that inspire us and provide us with reference points for our own work have a very different tonal quality. And while many video editors now offer 'old film' filters, they usually amount to little more than a sepia tone and superimposed scratches or dust. We are not trying to ruin the pristine look of our movies nor simulate the effect of badly treated archive footage, however, we just want some of the richness of colour and tone that we are used to seeing in big commercial DVDs. The gap between the look of big Hollywood releases and your own efforts goes beyond the medium itself or the cost of equipment. Of course, film has a much greater contrast range and looks inherently different from video, and the big production companies are using much more expensive equipment – their lenses alone can often cost more than a modest mortgage. Most important, however, is the fact that commercial movies are lit very carefully by experienced professionals who understand the contrast range of their medium and can place shadows and highlights in exactly the right places. And when it is all shot and edited, there is a careful grading process in which warmth, tone and contrast are tweaked to evoke the right 'feel' for the on-screen narrative. It is this grading stage that is overlooked by most low-budget and home video makers.

Red Giant's Magic Bullet filters go beyond the scope of an editor's basic colour correction features, presenting users with a set of tools dedicated to the job of cinematic-style grading. There are many different flavours of Magic Bullet, each currently being sold as a plug-in for existing software. At the top level, there's

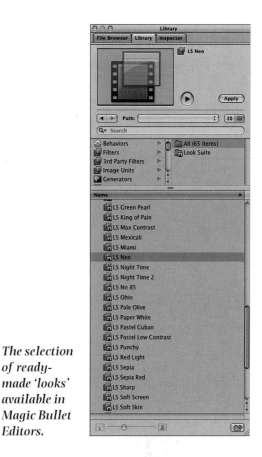

The selection of ready-made 'looks' available in Magic Bullet Editors.

the Magic Bullet Suite for Adobe's compositing program After Effects. This version is intended to help medium-budget movie makers prepare their finished video works for transfer to celluloid. More affordable and accessible to the mainstream, though, are the Magic Bullet Editors, which plug into Adobe Premiere Pro, Apple's Final Cut Pro, Sony Vegas and Apple's Motion software, as well as Avid's Xpress and Media Composer systems. As a plug-in, the workflow of Magic Bullet Editors can differ from one host application to the next, but all versions come with three different types of effect. There is a library of preset film 'looks', designed to match the tone and feel of popular movies and TV shows, such as *Three Kings*,

Monitor

(no clips) Effect Controls ×

Sequence 01 * PDRM0174.JPG

Video Effects

Look Suite

  ☑ Do Subject

  Subject

  Pre Saturation   −100.00
  Pre Gamma   0.00
  Pre Contrast   −10.00

  ☑ Do Lens

  Black Diffusion
  Grade   6.00
  Size   30.00
  Highlight Bias   0.00

  White Diffusion
  Grade   2.00
  Size   10.00
  Highlight Bias   80.00

  Gradient
  Grade   0.00
  Size   85.00
  Color
  Highlight Squelch   15.00
  Fade   50.00

  ☑ Do Camera

  Camera
  3-Strip Process   0.00
  Tint   0.00
  Tint Color
  Tint Black   0.00
  Tint Black Color
  Tint Black Thre...   30.00

  ☑ Do Post

  Post
  Warm/Cool   0.00
  Warm/Cool Hue   0.00
  Post Gamma   0.00
  Post Contrast   60.00
  Post Saturation   0.00

00:00:06:24

The custom controls for Magic Bullet's film-look filters are extensive.

Saving Private Ryan and CSI. There is also a hands-on editor, allowing every parameter of the grading process to be tweaked.

Third on the list of effects is Misfire – a selection of grain, gunk and scratch filters, if you really do want your digital video to look like an old cine reel. The presets are self-explanatory and have names that give a reasonable idea what to expect. 'Bistro', for example, simulates the soft warm grading for *Amelie*, while 'Neo' delivers the hard, green, high-contrast look of *The Matrix*. Other presets simply emulate the visual quality of film stocks or processing techniques – Colour Reversal and Buffalo give the oversaturated and high-contrast feel of the colour reversal film stock used in slide and 8mm cine film, and Bleach Bypass simulates the desaturated and contrasty effect of leaving silver in a final cinematic print – a technique exploited by gritty productions such as *21 Grams*, *Seven*, and *Saving Private Ryan*.

Hands-on editors divide grading into four steps – Subject, Lens, Camera and Post. Subject allows users to alter brightness, saturation and gamma before applying the Magic Bullet effects, which is useful, as we found that presets often resulted in very dark results if the source image was already high-contrast or highly saturated. Lens options simulate the effect of optical filters such as diffusers and graduated tints. Camera controls help to simulate three-strip Technicolor tones, as well as apply and control coloured tints to the overall image or shadows. Post options allow gamma, saturation and contrast to be replaced if they were altered in the Subject settings, and also give the means to warm or cool the final result. An attractive feature for anyone working in the compositing world of special effects is a de-artefact filter which removes blocky compression artefacts from DV and HDV footage – it is a blessing when working with greenscreens and trying to achieve a smooth, seamless key (more on keying later in this chapter). Misfire effects are fun, but not as immediately useful for day-to-day work. There are flicker effects, vignettes, gate weaves, dust, funk, splotches, grain and – of course – scratches, from tiny micro scratches to deeply scored trenches.

TITLING

Titling is the process of adding text to the video image. These titles can be static or animated and be placed over plain backgrounds or actual video footage. Titles serve a variety of purposes, from introductory slates to full rolling credits that scroll up the screen. They can also be presented as crawling text that moves right-to-left or left-to-right across the screen, and there is also very often a need to create text captions to provide additional information such as locations or the names of people in interviews.

Every video editor comes with its own titling tools. At the most basic level, you will be given the means to create static titles or animated ones that roll up the screen or crawl across it. There will also be a choice of typeface, usually giving access to any fonts installed on the system as well as options concerning text size and colour. In most cases, you will also be given the option of adding a border, shadow, sheen and even a 3D bevel effect to the text characters. For the purpose of creating information captions there may also be tools for adding geometric shapes or even graphics to the background in order to create even more separation between the text and the main movie.

The first of the two images presented here is Pinnacle's titling tool TitleDeko, which accompanies the entry-level Studio editor as well as the company's more advanced offering, Avid Liquid. It is a highly visual and intuitive interface with a good mix of galleries and preset options as well as more controlled tools for hands-on design. Next to that, you will see the titler that accompanies Adobe Premiere Pro. It does much the same job, but takes a more sober and formal approach. It arguably allows more control over the style of text, but is not nearly as tactile as Pinnacle's offering.

On the Mac platform, Final Cut Pro comes supplied with a feature-rich titling tool named LiveType, which provides an easy-to-use environment for creating animated titles from pre-made templates and tweaking whatever actions and parameters are necessary to fit the context of your movie.

Pinnacle's TitleDeko plug-in for Studio and Avid Liquid.

Adobe Premiere Pro's own built-in titling tool.

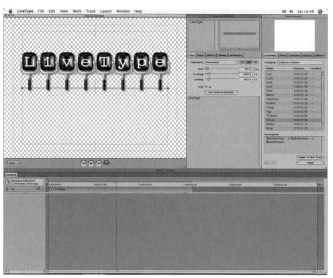

Apple's animated titling tool, LiveType.

ALPHA CHANNELS AND KEYING

There are many occasions during video editing when it is necessary to combine stills and video (we have already seen this at work in titling effects), but there is often also a need to create more complex compositing sequences with numerous video and graphic files. Many transparency effects of this type – especially those that overlay still images atop video – are achieved by the use of alpha channels. In the most commonly used editing applications alpha channels are created and identified automatically, making compositing a simple and immediate process. It is this automated application that makes titling so easy. For more complex effects, a higher degree of control is required, as is a basic understanding of how alpha channels work and how they are created.

An RGB colour image contains three colour channels, while a CMYK image has four. Each of the channels is an eight-bit, greyscale image that carries information about the distribution and intensity of a colour. These three or four basic channels are responsible for an image's appearance in a stand-alone state, but many graphics formats, such as PSD, PICT and TIF files, can carry up to twenty-four channels. The way in which channels work can be seen in well-featured image-editing programs such as Adobe's Photoshop. In this example, Photoshop is being paired with Adobe's Premiere Pro, but the principles are the same if you are using other well-specified image and video editors.

With an image loaded in Photoshop choose channels from the Window menu. This brings up a new window with three tabs, 'layers', 'channels' and 'paths'. The channels tab will be active and show each of the graphic's component colour channels. For RGB images there will be four rows showing, each with an eye icon at its left. There is one for each of the three colour channels (red, green, blue) and a combined option showing how they all work when combined. Highlight one of the single-colour channels to view it in isolation and you will see that the displayed picture becomes greyscale. Move from one channel to another and the image will change subtly to show how each colour is distributed. White areas denote a high colour density, while black areas receive none. Each channel supports 256 shades of grey, representing different degrees of colour density. Colour channels can be seen as a form of mask and alpha channels work in exactly the same manner but, in this case, they determine transparency for the image as a

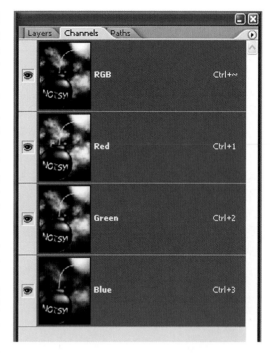

The separate colour channels of an RGB image, as displayed by Adobe Photoshop.

whole rather than the intensity of a particular colour channel. A white area will be opaque, while a black area in an alpha channel will represent complete transparency, and there are 256 levels of transparency in total.

With the channels palate open, click on the new channel icon next to the trash can at the bottom right. A new channel named alpha 1 will appear. It will be completely black, meaning that it is currently opaque. Double-clicking on this channel will call up a new dialogue box in which a name, opacity colour and standard level of transparency can be specified. The masking method is also set here – by default, colour in the channel is made to signify masked areas, but this can be changed to represent selected (visible) areas or spot colour as required. Spot colour channels are designed to identify the areas of an image that require special colours during printing and are not related to video applications. Pick a masking colour and choose a standard opacity level; 50 per cent is chosen as the default, but a setting of 100 per cent is possibly more appropriate for creating masks for video. Use Photoshop's draw tools to define its areas of transparency. When working with the default, colour indicates settings, the addition of white areas will result in complete transparency,

while shades of grey will result in a translucent effect. Click the eye icon on the other channels to see how your transparency effect will look.

Another approach is to right-click on an existing colour channel and select duplicate channel. This creates an alpha channel that will completely mask one of the colour channels. Further editing of the alpha layer with Photoshop's drawing tools or image filters can create some interesting textures and effects. Remember, too, that several alpha channels can be added to a single image, allowing one imported file to be created with several transparency masks available.

Having created a graphic with an alpha channel mask, save the file as either a Photoshop PSD document or in BMP format. Make sure that it has a 4×3 aspect ratio (or 16:9 if you are editing widescreen footage), so as not to be distorted when used in a video frame. Many editors – especially those at the entry level – will automatically detect alpha

Additional channels provide transparency masks for the image. These can be automatically detected and interpreted by many editing programs.

Adjusting the image size or canvas size before saving it is a good idea as it helps to prevent stretching or distortion when imported into the editing program.

Alpha transparency automatically applied in Pinnacle's TitleDeko.

channels and apply transparency automatically.

In the case of more advanced programs you might have to do some extra work to make the mask active. In Premiere, import the graphic file into a project. If importing a layered PSD file you might be asked whether you want to import a single layer, merged image or sequence. For the moment, choose merged layers. Drop your background video file on to video track 1 of Premiere's timeline, then add the graphic file directly above it on track 2, it will be automatically scaled to fit the frame. Highlight the graphic on the timeline and in the clip menu, choose video options and transparency.

In the 'key type' drop down list choose alpha channel. The image will become transparent in the dark areas of its alpha channel. If required, tick the reverse key box instead to make the dark areas opaque and the light areas transparent. Looking at Premiere's transparency settings, you will see more keying options than a simple 'alpha channel'. The first of these is 'black alpha matte' and

'white alpha matte'. As with normal alpha channel transparency, these effects rely on the image having a dedicated alpha channel. Choose 'black alpha matte' to make the black areas transparent, or 'white alpha matte' to make the areas of the image marked in white on the alpha channel transparent.

The mask used to identify transparent areas in an image file can also be referred to as a key. When working with two video clips it is often easier to key in irregular shapes using colour rather than a pre-made alpha channel

Import options for layered PSD images in Adobe Premiere Pro.

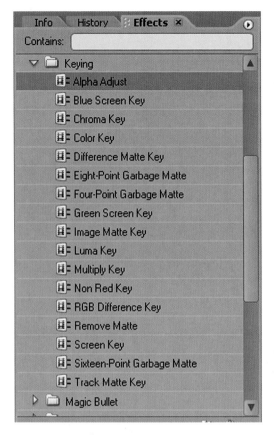

Premiere's key type options are typical of the features you will find in advanced editing programs.

Subjects framed against a green screen; the green background will later be replaced.

The same shot after keying.

mask. One of the best known techniques for superimposing videos is the blue screen. These have been used for decades in the cinema to provide a uniform coloured background that can be treated as transparent in post-production. Blue was chosen because there is little blue present in human skin tones. More recently, green screens have taken over in TV and video studios, since DV camcorders carry more detail in their green channels than in the red or blue ones. Using a green background colour helps to maximize edge detail and

ensure a clean key. The reflectivity of green paint is also slightly better than that of blue paint, which can make green screens easier to light. Be aware, though, that any green 'spill' caused by poor compositing will be far more obvious and off-putting to the viewer than the occasional flash of blue. Backlighting the subject with a slight yellow can be invaluable in preventing colour spills caused by the background's being reflected in white fabric. Also ensure that no blue (or green) clothing, props nor lighting are used on the subject when shooting bluescreen or greenscreen effects since these will appear transparent too when the images are composited. The same applies to green items when working with green screens.

A chroma key background must be uniformly lit and never underexposed. An easy option would be to light everything direct from the front, but that also kills any dramatic tone you want to set with your actors. If the blue screen is large and studio space is ample, it will help to put some distance between actor and screen, allowing better control over independent lighting of background and foreground. For more confined spaces, some video makers choose to have their screens backlit. Anyone serious about bluescreen

effects should invest in a video waveform monitor too, just to make sure that there is absolutely no fall-off in the background lighting that could interfere with compositing later in the process. Of course, the best method of attaining perfectly even lighting is to have the sun light the set for you.

Replacing the blue or green background with different footage is a snap in most editing programs. Even affordable, beginner's software such as Ulead VideoStudio or Pinnacle Studio now offers bluescreen and greenscreen keying effects. In most cases all you need to do is to tell the software that you are using a bluescreen or greenscreen effect and the rest is automated. With Premiere,

A chromaflex light ring and screen help to ensure that the background is uniformly lit regardless of the lighting used on the subject in the foreground. A green or a blue light ring surrounds the camcorder lens and its light is reflected back from a ground glass screen behind the subject. Results are typically excellent, but care must be taken to ensure that no reflective surfaces are present in-shot that might reflect the ring itself.

Simple chromakey in Ulead VideoStudio.

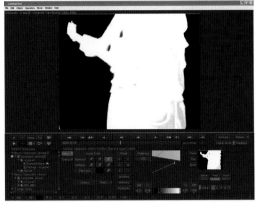

More advanced chromakey tools in Discreet's Combustion software.

keying options are chosen from the effects palette and more control options are provided to help to ensure a smoother key. Moving even further up the effects ladder to dedicated compositing editors, you will find that hands-on controls become much more extensive. In the example pictured here, we see greenscreen keying in Discreet's Combustion – notice the huge array of settings for defining key colours and fine-tuning transparency masks.

Rendering and Real-time

A big consideration to take on board when playing with special effects is the need to render footage once they are applied. Rendering is a process whereby the affected video footage is remade as a single, contiguous clip. It is all seamlessly done to ensure smooth playback from the timeline, but can often take time to process and rob you of valuable editing time. As computers become faster and more powerful, so rendering becomes less of an inconvenience. Extensive colour correction and compositing are almost certain to require rendering time, and, in these cases, your sanity can be saved in one of two ways – first, save all effects work until the very last stage of the editing so that you can leave the computer rendering overnight if need be; alternatively, invest in a faster computer or some dedicated, real-time hardware to enable you to view effects immediately. Real-time solutions are a great bonus for feeding movies to TV sets for high-quality viewing, but many of them (especially those working with the high-definition HDV formats) still require formal rendering to take place before the project can be exported digitally via FireWire. Real-time boards are useful but pricey. If you are an established jobbing video maker and earning money from video it is a wise investment. If you are a hobbyist, think carefully before spending your money.

8 FINISHING OFF

EXPORTING AND SHARING PROJECTS

The whole purpose of making a movie is to have it seen by an audience. Inviting all potential viewers round to your house to watch it in your video editing system isn't the most practical way of having it screened, so we need a way to get it off the computer and on display. Video export options are now quite numerous, with digital and analogue video tapes, standard and high-definition discs and internet-based distribution methods. Each distribution method brings with it its own set of hurdles and choices and the subject is too vast to be dealt with adequately in one chapter of such a work as this. What follows is a basic overview of the available options.

EXPORT TO TAPE

Having completed a video edit, mixed the sound, added effects and graded colour, your first priority should be to make a high-quality backup of the finished movie. If you are working digitally in a MiniDV, Digital8 or HDV format the best quality you can get is to send it back to digital tape via FireWire. For DV projects, an editing system may need to render all transitions and filters before export, but the process is an efficient one – rendering only material to which effects have been applied. Doing this saves time and hard drive space. What is more, some of the more advanced real-time editing boards allow the immediate output of video effects via FireWire without the need for rendering. They are convenient

VHS is still with us and recorders are inexpensive, but demand for VHS tapes is now extremely small.

but expensive, and also have their limitations. The more effects you add to a particular clip, the more likely it is to need rendering. Whether you are working with or without this kind of hardware acceleration, however, the speed of your computer system will have an immense impact on rendering speeds.

For HDV edits, things are currently slower and more cumbersome. At the time of writing, it is still necessary for completed HDV projects to be saved to the hard drive as a complete, self-contained video file before being sent to tape via FireWire. I am hopeful that this will change

in the near future as HDV technology becomes more mature, but for the time being it is important to leave enough hard drive space on the system to accommodate a full copy of the finished movie for export.

When it comes to sending the project back to digital tape, however, the software will take on the job of controlling the connected camcorder or deck, automatically starting and stopping the tape rather than relying on manual intervention. If you are using a connected camcorder as the recording device it is important to make sure that the machine

World Video Standards

If you are preparing video tapes or DVDs for other people, keep in mind the countries in which they are living. Different countries have different television standards and your tapes and discs will need to comply with their technical specifications if they're going to play properly. For DVD distribution, you need worry about only two standards – PAL and NTSC. PAL is the standard used in the United Kingdom, mainland Europe, Australia and New Zealand. NTSC, on the other hand, is used in the USA, Canada and Japan. Without going into great technical detail at this point, the two standards have different frame resolutions and different frame rates. Players and TV sets in PAL countries often do a reasonable job of interpreting and playing NTSC discs, although you will still find the occasional player or TV set that cannot handle the foreign signal, choosing to display it as a black and white image. NTSC countries have a much harder time handling PAL video, and, while it is becoming common practice for niche-market distributors to release NTSC discs worldwide, it is considered to be a big mistake to try launching PAL media into the North American market. To change the video standard

is a reasonably simple process if you have the right tools. Canopus's media encoding programs ProCoder and ProCoder Express do excellent jobs of converting between the PAL and the NTSC format and will also encode to DVD-compliant MPEG files at the same time. Remember, though, that if two different video standards are required you will have to author two different versions of the DVD – PAL and NTSC videos cannot be included on a single DVD disc. But things are more difficult when preparing VHS tapes for clients abroad. Not only do you need a specialized deck to make the actual recording, but there are far more TV standards to worry about – different variations of PAL and NTSC, as well as the French SECAM standard and the African MESECAM format. Specialized electronics dealers will be able to sell you a multi-standard converter and recorder for making these VHS conversions, and they are remarkably easy to use but they are much more expensive than standard VHS recorders and, with the increasing lack of interest in VHS, I doubt that it would see enough use to justify the cost.

Set-top DVD recorders make it quick and easy to export movies to disc but lack versatility in terms of menu design and content structure.

has a working DV input (many camcorders in the EU have had their video inputs disabled so as to avoid excessive import duty). In addition, it is worth remembering that camcorders that take MiniDV cassettes will offer a maximum running time of only 60 minutes. While it is true that long-play recording gives a 90-minute recording time, it is not a recommended practice for masters and backups as slower tape speeds are more prone to errors. Similarly, I would not advise the use of 80-minute MiniDV cassettes as they use a thinner tape stock which is more prone to faults. If you regularly make movies with durations exceeding an hour, I would advise your investing in a deck that supports full-size DV tapes (supporting recording times of up to 4 hours) or divide the project into sub-60 minute chunks before export so that it can span several tapes.

Analogue tape is on its way out and the demand for VHS copies of your work is sure to be minimal. Most people today have DVD players and prefer the improved quality and convenience that the format offers. There will be the rare occasion that someone asks you for a VHS tape, however, and this is easily made by connecting the camcorder or deck to a VHS deck and making a direct dub from the master.

Alternatively, if you haven't made a digital master tape yet, route the computer's FireWire feed through a camcorder or deck's analogue outputs (if the camcorder allows this) and dub direct to VHS from the editor's timeline.

CREATING DVDs

DVD has become the video format of choice in the mainstream video market. Rentals and sales of commercial movies are now almost all DVD format today and VHS is quickly being phased out of shops and video clubs. The appeal of DVD is clear – picture quality is significantly better than that of VHS and the format supports surround-sound, multiple video angles and chapter points to allow immediate access to specific scenes. Menu structures also allow more than one video to be accessed in an instant and make it easy to navigate extra features such as photo slideshows. A single movie on DVD can contain up to eight audio tracks and thirty-two subtitle feeds, allowing producers to make one disc to meet the demands of different territories with different languages. The discs are small and lightweight and require less room to store. DVD is better than VHS in all respects, and the tools for making them are

In order to create a DVD on your computer, it will need to be fitted with a DVD burner. Many new computers have these as standard. If yours doesn't they're inexpensive and easy to install.

Recordable DVD Standards

One of the biggest confusions with DVD burners centres around the different types of disc on offer. When the market was fresh, drive manufacturers were keen to support only one of the two competing DVD recordable formats available – DVD-R/-RW and DVD+R/+RW (very well, DVD-RAM had been around for a few years, but had proved to be unsuitable for video distribution and playback in set-top players). In fact, DVD-R and DVD+R discs were found to be incompatible with some of the earliest set-top DVD players then available, but that is now largely a historical issue. In recent years, any confusion over recordable DVD formats has been rendered moot, thanks to a move by most drive manufacturers to support both camps rather than subscribe to just one. Care should be taken when choosing blank discs though, always opt for known, brand-name discs, regardless of the cost, as many unbranded budget offerings are B-grade stock and may be prone to faults. For video distribution, write-once DVD-R or DVD+R discs should be used as they are more robust than re-recordable media. Where possible, choose the slowest burning speeds your software will allow. Doing so will take longer, but help to reduce the risk of burning errors. Also steer clear of double-layer DVD+R discs for video distribution; the high capacity discs might seem ideal for putting long projects such as weddings and concerts on to a single disc, but few authoring programs offer control over the placement of layer breaks, meaning that playback could pause at the most inconvenient places as the player switches layers. On top of this, many set top players have great difficulty navigating layer breaks – the pause in playback that would last a second with commercially-pressed, dual-layer discs might take a minute or more with home-burned DL media.

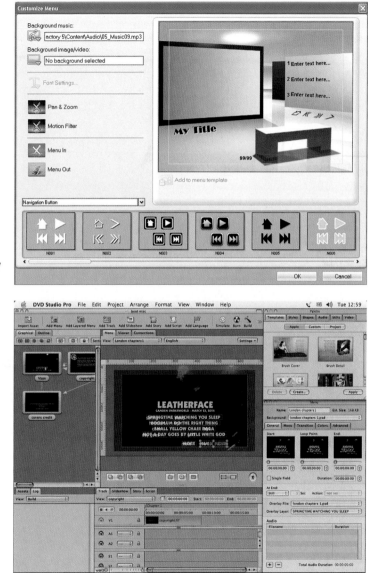

Entry-level DVD authoring software from Ulead.

Apple's DVD Studio Pro is a much more advanced authoring tool.

becoming cheaper and more accessible all the time.

DVD burners for desktop computers can be bought for well under £50, and DVD authoring programs (the software required to format media, generate menus and burn discs) are outrageously cheap, often being included as an afterthought with general music and data disc-burning software such as Ahead's Nero or Roxio's Easy Media Creator. Entry-level authoring tools will allow you to create discs with main menus and scene selection menus giving access to individual chapters within a movie. They will also support the addition of slideshows and allow menus to be static or video-based.

Advanced features, such as multiple audio tracks and subtitles, require slightly pricier, dedicated DVD-authoring tools. Offerings such as Ulead's DVD Workshop and Adobe's Encore DVD serve the purpose well at relatively keen prices and also support professional mastering to digital linear tape – the preferred format for mass factory replication. In terms of value, however, there is still little in the affordable Windows software market that can compete with Apple's Mac-based DVD Studio Pro, which also supports multi-angle video and complex scripting tools. Sony's DVD Architect is making great strides forward, but the Windows market is still struggling to keep up at the prosumer level.

VIDEO ON THE INTERNET

The days of slow internet connections and text-driven websites are over. Broadband is now readily available to almost anyone with a

---

High-definition Video Discs

As this book goes to press, high-definition video is slowly making its way into people's homes. In simple terms, high definition delivers greater picture resolution than we currently see on standard broadcast or formats such as DVD. But it isn't a new idea – high-definition video has been around for decades, but has largely been touted as a low(er)-cost alternative to film when it comes to shooting commercial movies. Attempts have been made over that time to push high-definition video into the broadcast market, but that has, until now, been blocked by broadcasters who are keener on squeezing more channels into their allotted bandwidth and maximizing advertising revenue than pouring time, bandwidth and resources into improved picture quality. Large LCD and plasma screen TV sets are now affordable and becoming immensely popular and their ability to display much higher resolutions is whetting people's appetites for high-definition content. Digital broadcasters are relenting and taking steps to begin transmitting high-definition channels, while movie distributors are embracing new disc technology to put their movies out in high-definition formats. On one side, it can be seen as a licence to print money, as keen viewers will need to invest in a new player and buy all their favourite movies again in the new disc format. On the other, the market is moving forward on very shaky ground, thanks to a format war between companies who want to hold the definitive patent on the next generation of digital video disc. In one camp we have HD DVD (high-definition DVD) and, in the other, there is Blu-Ray. On paper, Blu-Ray would appear to have the advantage over its competitor, thanks to the discs' larger capacity, making it capable of holding more material at higher data rates than HD DVD. Early comparisons, however, have indicated that HD DVD uses a more efficient method of video compression, allowing it to deliver very high quality video at a smaller bandwidth. The arguments are set to run and run (as they once did for VHS and Betamax video tape formats). In the meantime, disc burners for both formats are slowly being released into the consumer market (albeit at debilitating prices) and even some entry-level DVD authoring programs are racing ahead to support the new standards. The technology is still in its infancy though and it would be inappropriate for me to go into more depth than this now.

computer, and, when you factor in the savings made on phone calls, it can actually work out cheaper than using a dial-up service. Internet connection speeds of up to 8Mb are now common, and that means quicker download times and access to rich online media, including music and video. Video can be served online in one of three ways: streaming, download or progressive download. Downloadable video is possibly the most straightforward method of delivering work. Viewers must copy the whole video file on to their computer hard drives before it can be played. There is a disadvantage in that they must wait for the whole to download before it can be viewed, but you won't encounter the same fluctuations in speed and quality that can beset real-time online viewing, especially if the site is busy or the viewer's connection is slow or unstable. Apple's QuickTime format with an MPEG-4 codec is well suited to video downloads, but the most commonly used compression method – especially for file sharing online – uses a DivX codec (itself a derivative of MPEG-4) in an AVI wrapper.

Streaming video serves footage online in real time. Media is brought into the computer in packets which are displayed and removed from the system immediately after playback. In this way, content providers and distributors have some control over their work, knowing that it is very difficult (but not impossible) to copy and pirate. It also means that viewers can watch movies straightaway without having to wait for the entirety to download. Another advantage with streaming video is that footage can be encoded at several data rates, all wrapped in the same file. Doing this allows the quality to be increased or degraded depending on the connection speed to ensure smooth, uninterrupted playback. Dedicated streaming servers are required for this and services are not cheap. Prices are much keener now than they were five years ago, however, and anyone eager to protect his work and give a slick, fuss-free viewing interface on their websites is well advised to look into it. The most commonly used streaming video codecs at the time of writing are Windows Media and RealVideo, although Macromedia Flash Video is becoming more attractive due to the way in which it can be integrated into media-rich Flash websites.

Progressive downloads fall somewhere

Streaming or progressive download media can be made to play in a software player, or embedded into the website's design.

between the two methods mentioned above. A progressive download video is one that must be brought on to the viewer's computer but which can begin playing before the whole file has been downloaded. These videos don't require dedicated streaming servers, making it cheaper to put them online, but they can be saved to the viewer's computer and distributed freely on the web. There is also no support for multiple bandwidth playback, such as you have with streaming video. Progressive download is where Apple's QuickTime format comes into its own and is possibly the most commonly used method of delivery in this field.

An important consideration to take on board before you try to publish video online is that video can be very data-heavy and web hosts often assign limits to the amount of data transfer you can have per month. Even those hosts that claim not to have such a transfer limit may include a 'no downloads' clause in their user contracts, allowing them to close your account and switch off your site if the traffic becomes too high. With this in mind, many video enthusiasts choose to publish their videos through community sites such as MySpace and YouTube instead. It is not as elegant as having work embedded in your own web page, but these are services that actively encourage you to put media online.

MULTIMEDIA

The uses of video in multimedia have traditionally centred around the creation of interactive presentations on CD-ROM or kiosk-style information displays. Music publishers are now prone to including videos on enhanced audio CDs for playback on computers and a small market has also appeared for the technically savvy to create CD-ROM business cards featuring short multimedia presentations. These are exciting ways to use video, and in recent years it has become easier to create presentations that run on the majority of home and office computers. There is a much freer choice of video format for this sort of work too, limited only to the sorts of player or codec that the viewer is likely to have on his system. Or if you are creating a self-contained presentation in Director or Flash, that's even less of a problem. MPEG-1 and even MPEG-2 are often considered appropriate for these jobs, as is QuickTime.

Beyond the scope of promotional presentations and CD-ROMs, video is now being used on small, handheld media players – the type that you would normally associate only with music playback. The new breed of MP3 and digital audio players are now being made with small but high quality LCD screens and the ability to play video material in various formats. Even mobile phones now double as audio and video players. In these cases, however, it is understood that most of your media will be acquired online, and so the supported formats are typically DivX, Windows Media and QuickTime.

9 IN THE REAL WORLD

Having learned the root techniques of video editing, it is now time to apply them to actual situations. As with most artistic methods, there is no hard and fast, right or wrong way to do things, but what follows is an overview of methods and approaches that suit the most common methods of editing for different types of project. While it is my aim to show the real-world applications of editing techniques, much of what follows depends on the assumption that footage has been shot in a particular way. The bottom line to remember when reading this is that the correct approach to editing depends on the footage you are working with and the ways in which it has been acquired. Unorthodox shooting techniques will require unorthodox editing techniques – that's not a bad thing though. Experimentation and lateral thinking are essential for the development of film and video, but what it also shows is that there is no one-size-fits-all recipe for a first-rate motion picture.

MAKING HOME MOVIES WATCHABLE

By home movies, I refer to family videos – candid moments caught on your camcorder of a holiday, birthday or day out with family and friends. These are the situations where little real care and attention are given to the task of shooting video – nor should they be. The more time you spend worrying about the composition, sound quality and lighting, the less time you will have to enjoy yourself and actually be part of the event. This is truly spontaneous material and in that sense it is possibly the most honest footage that you're likely to work with. *Digital Camcorder* gave some practical advice for shooting informal video. These are solid tips and should be carefully considered, but it is not a crime, however, to point and shoot in these situations. Even I do it.

Many camcorder makers work under the assumption that the people that use their machines have no interest in editing the footage they shoot, and that is certainly the thinking behind many DVD-based camcorders, which are designed to give you a watchable disc immediately but don't lend themselves well to editing. It is very easy, though, to shoot too much material, having the camcorder running just in case something interesting happens. It is also likely that you might mean well but shoot too little, ending with a series of disjointed moments that don't make any real sense either on their own or in sequence. Editing these movies into a coherent form is only polite if you plan to share them, and it is also good practice if you ever plan to revisit these memories in future.

At the very least, home movies should not

115

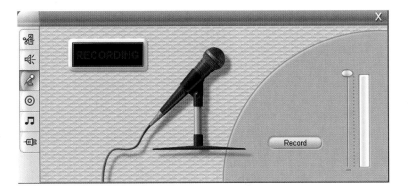

Most video editing programs at all levels – such as Pinnacle Studio, pictured here – provide the means to record commentary direct to the timeline.

be allowed to drag. One of the common mistakes with family footage is that shots linger far too long and whatever narrative is being portrayed drags on and on, ad nauseam. Identify the active moments of your footage. When does the interesting stuff happen? Isolate that footage and ditch the rest. Remember that your raw footage will always be there on tape unless you actively choose to erase it, so don't feel too protective about every individual frame. Remove the footage that serves no purpose and compile your edit purely of useful material that serves to tell a story. If you have been in the habit of narrating to the camcorder as you shoot, make use of that soundtrack. The visuals may be dull and wobbly, but you can put more useful illustrative footage and graphics on top of it to give a more solid form and structure (see insert editing in the chapter on 'Simple Structure'). Narration is also a good tool if you are stuck with lots of short moments that don't add up to much. Most editing programs allow commentary to be recorded direct to the timeline during playback, and you will find this a great tool for tying numerous loose ends into a single, coherent strand. Many hobbyists take a different route by cutting cute montage sequences to music – that is an acceptable way out if you can keep it interesting, but, after a minute or two, such sequences can become

tiresome. Montages are a good way of using abstract shots, but try to keep them short and use them as links between narrative pieces rather than making them the movie itself.

CUTTING DRAMA – FLUIDITY AND RHYTHM

Drama is unlike any other form of video in that its edits are typically constructed from several takes of the same event, shot from different angles and at different times. While live events such as concerts are shot as an entire show on three or more camcorders, a drama shoot will normally involve multiple performances captured by only one camera. Synchronization and continuity between takes will never be entirely accurate, and it may be the case that only certain parts of a scene are shot from certain angles. As editor, your job is to assemble these pieces into a fluid and coherent sequence that appears to run in real time. This is where sound becomes such a helpful tool – if a movie's soundtrack is continuous and smooth, viewers will be more inclined to believe in the immediacy of a scene, rather than seeing it as a number of shots that were taken over many hours or even days. Use techniques such as insert edits and audio splitting to give a greater independence to the movie's soundtrack, rather than simply cutting sound and picture at the same time. These steps will also encourage the use of cutaway details and reaction shots, all of which are invaluable in building drama and holding the viewer's attention. Remember that drama is just as much about reaction as it is about action, so allow your edit to give time and credence to the way in which characters respond to events. On top of the insert and split edits, carefully created spot effects will help to add depth to the on-screen environment, and the appropriate use of ambient wild tracks will top the whole thing off and give the entire scene a sense of organic cohesion.

Pacing is also an important consideration, however. The speed at which you cut, the variety of different shots you use and the variety in shot length will all contribute to the general feel of a piece. Try watching a modern action thriller with a stopwatch in your hand and time the length of individual shots during the big chase sequences – you will be lucky if you find two cuts spaced more than 5 seconds apart. Fast cutting adds an immense tension to a sequence, and you might also notice that the last two decades have seen a marked increase in the use of close-ups when shooting and editing action scenes. Compare the dizzying car chase sequences in *The Hitcher* with those of the earliest James Bond movies and you will see a much more active use of character close-ups (primarily reaction shots) and technical details (wheels, pedals and rear-view mirrors, for example). Establishing wide shots are still present in *The Hitcher* but they don't dominate the scenes as they did in the early Bond efforts. That is not to say that fast editing is an absolute requirement for building excitement. A fixed, long, unblinking shot can yield a lot of menace if played correctly – see the last 15 minutes of John Carpenter's *Halloween* for a prime example. It all depends on the tone you are trying to set.

Think carefully about the rhythm of your editing, making sure that it matches the mood of your scene and accentuates the emotive content rather than detracting from it. Some professional directors and editors are in the habit of laying down pieces of music that they feel match the emotions they are trying to convey before they start editing. The piece is cut to the music, taking guidance from its flow and rhythm, before the music itself is removed, allowing the scene to stand on its own. That might seem like an odd idea, but you would be amazed how often it works.

Among your cutaways and establishing shots look for good, strong footage that can serve as a lead-in to new scenes. Often a straight cut from one scene to the next can be disorientating, and placing a dissolve or wipe at every scene change might be seen as lazy editing. Strong graphic details of props or settings, on the other hand, let us know that we have moved on somewhere else, as do sprawling landscapes or cityscapes – notice the persistent use of night-time cityscape shots in TV shows such as *CSI*. They simply remove the viewers from the current action and allow them to draw breath before the next scene begins. Alternatively, some editors choose to drop their audiences straight into the next scene using a long split edit – starting the new dialogue well in advance before giving a straight cut to the next shot. It is not uncommon to deliver the whole first line before cutting to a reaction shot of the person being spoken to. But however you choose to play it, keep the move between scenes more dynamic and attention-grabbing than the cuts between shots within a scene.

CUTTING DOCUMENTARY

Documentary is a very loose genre, encompassing an enormous number of shooting styles. Some documentaries are composed entirely of archive footage such as newsreels, possibly with a voice-over to give it a coherent narrative thread. Others contain all newly-shot material, as a story is followed from beginning to end, and a growing number, such as the recent movies of Michael Moore, use a combination of archive material and original footage such as interviews. Documentaries can also be shot on-the-hoof, trying to catch a moment as and when it happens, or more controlled and – in some cases – completely staged or reconstructed.

Unlike most dramas, however, many documentaries find their form, structure and voice at the editing stage. It is easily possible to come out of a documentary shoot with a very different story from the one you intended to capture – and that is one of the many things that makes documentary so exciting. And while there's no right and wrong way to approach documentary, I can convey some practical suggestions to strengthen the movie, streamlining the editing process and helping to protect your sanity.

First, get to know your footage well before editing begins. If necessary, make a quick duplicate of material to VHS tape or, better yet, to DVD with chapter markers automatically placed at every 'scene break' (a point at which recording was paused and resumed). Watch the footage through in its entirety. Log the tape contents and make notes of shots that are good visually, good in audio content or generally useless. If your editing program supports the creation and management of bins, organize all useful clips into their own folders based on date, location or scene, whatever structure you plan to use when editing. If possible, subdivide clips further with categories such as interviews or cutaways. Knowing where shots are will save you a huge amount of time as the edit progresses.

Decide on a voice and stay with it throughout. All documentaries are biased; whether we intend it or not, they convey somebody's reality. Decide in advance whose world you are representing. Is this your personal perception of the world? Or will your viewers be seeing it through the eyes and opinions of somebody else? Don't try to avoid the issue, any conscious attempt to make a documentary 'unbiased' is likely to leave it confused at best and just plain boring at worst.

As with identifying the voice of this piece, also decide on a storytelling method and keep it going for the entire project. Some documentaries make excessive use of voice-over to provide the movie with a backbone. Others have no audio commentary at all and leave the on-screen subjects to lead us through the movie's narrative. In many cases these choices are made from necessity and you will find that, the more thoroughly you cover a subject and events, the less need there is to explain it with voice-over. However, if the movie as-is does not convey the opinions you want expressed then audio commentary is an effective way to intervene. Either way, don't flit between methods – don't add commentary to one scene in the middle of a movie if the rest of it is getting by nicely without. Similarly, if viewers have become used to the idea of a narrator's telling them the plot as it unfolds then don't suddenly take that lifeline away and expect them to follow the narrative unaided.

CUTTING MONTAGE TO MUSIC

One of the most common uses for video editors among novices is the creation of montage sequences, cut to music. Often, these projects come about because the user can't find a strong enough narrative thread with which to tie his material together. As we have seen, the addition of a simple audio commentary would make a world of difference and provide a much-needed context and significance to shots that once seemed random and disjointed. It is much more common, though, for novices to

Automated montage creation in Pinnacle Studio.

strip off the original audio and assemble their clips to a piece of their favourite music.

These montage sequences are also common in some professional projects. They are seen in many dramas, documentaries and advertisements and serve a variety of purposes, from illustrating the passing of time to caressing the contours of an attractive new product and making you want to buy it.

The editing of montages can be as thoughtful or as mindless a process as you want it to be. In fact, many beginners' editing programs provide simple wizards that will take in your footage and churn out a montage automatically, with little or no input from the user. In the better examples, however, these automated sequences can then be opened in the program's main editor and customized to your own liking.

If you are cutting montage sequences without the use of an automated wizard place the background music on the timeline first and enable visible waveforms, providing a visual representation of the music and allowing you to cut to the beat accurately. From there, the process is a simple task of trimming and assembling footage and arranging it on the timeline with simple assemble edits. The original audio from the camcorder footage can be muted so as not to distract from the music and cutting rhythm.

Adobe's Premiere Pro has an ingenious set of tools for montage creation. First off, music is dropped on to an audio track and timeline markers are placed at each of the beats. This must be done manually, but it is as simple as hitting the '*' key on your keyboard. Do this in real time while playing the music or navigate visually using the timeline's waveform display.

Next, the project window is going to be used as a makeshift storyboard. Ensure that icon view is enabled at the top of the window and drag all the clips for your edit into a crude sequence, running from left to right, top to bottom. That done, select all clips and select Edit>Automate to sequence. The selected clips are then sent straight to the timeline and cut at the marker points you set earlier.

Despite being a colossal time saver, the job isn't quite finished yet. All clips begin on their

Markers placed on the timeline at potential cutting points; in this case we have been following the beat of the music.

first frame and you will probably want to alter their in and out points. Simple slip editing is a tactile and visual way of doing this. Select the slip tool, click the required clip and drag to the left or right.

The automate to sequence command.

MULTI-CAMERA EDITING

Projects shot in real time with two or more camcorders require a quite different editing approach. Rather than assemble sequences clip by clip, it is more productive to synchronize all media on parallel timeline tracks and edit out the shots that are not required. In order to do this, it is important that all the camcorders have been left running continuously throughout the duration of the show you are recording. If you haven't paused the camcorders at any time, then shots will need to be synched up only once.

Synchronization is easiest if you have a visual cue for each camcorder which can be used as reference on the timeline. An ideal tool is the traditional clapperboard, which gives a visual reference (the moment at which the board closes), coupled with an audible cue (the actual clapping sound). This is essential for film production, where picture and sound are recorded on separate devices and need to be synched up later. A clapperboard might be impracticable, however, if you can't get a close-up of the board on every camcorder at once or if you just do not have the opportunity

120

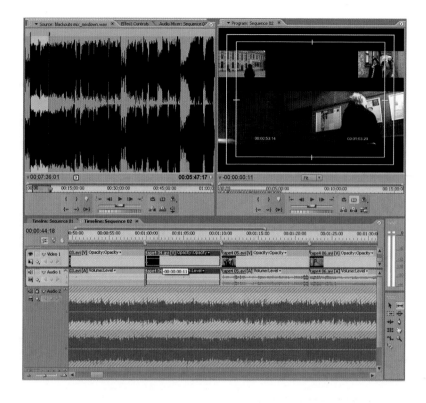

Slip editing in Premiere Pro keeps the cutting points at the same place on the timeline, but alters the start and end points of a specific clip.

of running into frame with one. A good alternative for video production is a camera flash. Flashes will be detected on only one frame of video (often only one field), and if they have been picked up by all camcorders you are left with a precise and foolproof reference for cuing up your footage.

Until recently, the only sensible way to edit multi-camera material in an editing program was to place all clips on parallel tracks, neatly synchronized so that your cue points mentioned earlier fall at exactly the same frame of the timeline. That done, cuts are made by slicing through the group of clips and all but the clip you want is either deleted or disabled. Not all editing programs will allow you to simply disable an unwanted clip, but if yours does, I would recommend this over deletion since it makes life easier, should you

change your mind about editing decisions at a later date.

Naturally, this kind of editing can be done only with more advanced software which supports multiple video tracks and also provides the tools to slice through multiple clips and toggle entire tracks on and off in order to see what's happening on each. Most beginners' programs are quite unsuitable for the job.

Remember too, that sound needs to be carefully considered. If audio has been recorded on a separate device, synching it up to the picture should be relatively easy by comparing the peaks and troughs of visible waveforms between the 'good' sound and the rough audio captured by the camcorder. If, on the other hand, one of the camcorders was used with good mics to collect sound as well as

121

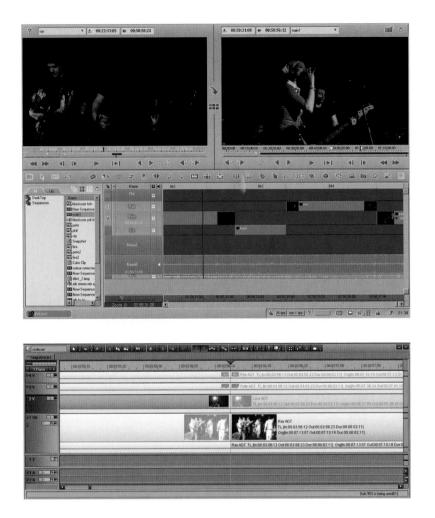

*The difference between deleted and disabled clips when editing a
multi-camera project.*

picture, then your audio is already in sync, but
needs to be protected on the timeline. Before
editing, make sure to lock the appropriate
audio track, making it immune to slicing,
deletion or disable commands. Also take a
moment to mute any unwanted sound that
came in with the other camcorder feeds.

More recent versions of programs such as
Final Cut Pro, Avid Liquid, Adobe Premiere
and Grass Valley Edius Pro have come with
dedicated multi-camera editing tools. Clips are
synchronized as before, but users are then
given an interface featuring multiple video
monitors displaying each of the source video
feeds, and a second monitor showing the
currently selected clip and edit in progress. The
number of supported video feeds depends on
the software you choose and may vary

between a maximum of four and sixteen, but be aware that the more video feeds are displayed at one time the smaller each one will appear on screen, and the more playback performance will be compromised. Depending on the power of the host system, a sixteen-feed display might suffer a badly reduced frame rate and be seen as staccato and jerky, making it more difficult to make detailed decisions.

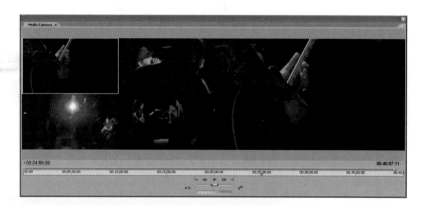

Multi-camera editors provided by Premiere Pro and Edius Pro.

GLOSSARY

Animation
Creating the illusion of movement from a selection of still images. Typically achieved by shooting one frame of film at a time and moving the subject between shots, although much animation is now done on computers.

Aspect ratio
The shape of a video frame. A 4:3 aspect ratio means that the picture is four units wide and three units high. Modern TV sets and broadcast images now have a widescreen aspect ratio of 16:9 and many cinematic movies are wider still.

AVI
Audio Video Interleaved – the most common file extension for video files on Windows PCs. The AVI extension tells us nothing about the method of compression used, however.

Batch capture
Capture of two or more logged video clips to a computer's hard drive in a single, unattended session.

Bit
The smallest unit of digital information, and often referred to as ones and zeros. Eight bits make up a byte and this is the unit used to measure memory and colour depth.

Bitmap
A digital image composed of a mosaic of coloured or greyscale pixels. Bitmap is also the name given to Windows graphics saved in a DIB (Device Independent Bitmap) format with a BMP file extension.

Bitrate
The amount of data that is read or written in a single second. DV Video, for example, has a bitrate (or data rate) of 3.6MBytes per second.

Byte
Eight bits. Note: there are 1,000 bytes to a kilobyte and 1,000 kilobytes to a megabyte.

CD-ROM
A CD disc containing computer data rather than music.

Chrominance
Colour information in a video signal.

Codec
Compressor/decompressor. The key to a video file's compression. A video player will need to have access to a particular codec in order to play files that were compressed with it.

Component video (YUV)
Division of colour and luminance with a video signal, where Y is the luminance (black and white) information, while U and V are colour channels. A component video feed is one of the highest quality methods of sending analogue video signals.

Composite video
Video signal in which all colour and luminance information is passed in a single channel. All VHS decks have composite video inputs and outputs.

Compositing
Combination of two or more video or graphic elements

124

in a single video frame. Compositing is used to create special effects for advertisements and feature films, for example, where subjects were shot at different times or some were created in computer software,

Compression
Reduction in the size of a digital file. Compression can apply to video, audio or graphics. Most methods are 'lossy', with information being discarded during compression.

Cut
A direct change from one shot to another without the use of a transition.

Deinterlacing
Combining fields of interlaced video to create a single, self-contained image. Of particular use when preparing video for projection, multimedia or online streaming.

Device control
Control of a camcorder's or VCR's playback functions direct from the computer's editing software.

Disc burning
Transferring data to a recordable disc.

Download
Acquiring data files from the internet and copying them to the computer's hard drive.

DV
Digital video, often specifically refers to Sony's DV and MiniDV.

DVCAM
Sony's professional version of DV. Video data is identical to DV, but tape runs faster and recordings are therefore more robust. DVCAM camcorders often have more professional features than their MiniDV counterparts, such as XLR audio sockets.

DVD
Digital Versatile Disc. A DVD disc can be DVD Video (for movies), DVD Audio (for high-definition music), or DVD-ROM for data.

DVD authoring
Formatting video, audio and photo slideshows on to DVD disc, usually with interactive menus and special features such as subtitles, alternative soundtracks or even multiple video angles.

DVD burner
A computer drive used to copy data to recordable DVD discs.

Edit decision list (EDL)
Text-based list of editing decisions used in a project. EDLs can be used to reassemble movies from source footage long after completion.

Filter
An effect used to alter the appearance of video or the sound of audio files.

FireWire
Apple's name for the IEEE1394 interface, and also referred to as iLink by Sony. It is essentially a communications port, used to connect devices such as camcorders, webcams, scanners and printers to computers.

GByte
1,024 megabytes.

HTML
Hypertext Markup Language. The scripting language used to author most of the internet's websites.

Interlacing
Division of a video frame into two interlaced fields. Most video shot for playback on TV is interlaced, with a slight time delay between the capture of each field. For that reason, interlaced video running at 50 or 60 fields per second can appear to have smoother motion than deinterlaced video running at 25 or 30 frames per second.

Keying
Compositing images or video clips by making a specific colour or tone transparent.

Linear editing
Assembly of video clips in sequence from beginning to end. Often done between two or more analogue video cassette recorders.

Luminance
The brightness information in a video signal. Luminance is often separated from colour information

for transport along S-Video and component video cables.

MPEG

Motion Pictures Expert Group. MPEG is a method of media compression. MPEG-1 video is used for VideoCDs, multimedia and some web applications. MPEG-2 is used for DVD, digital TV broadcast, hard drive-based camcorders and HDV camcorders. MPEG-4 is a high-quality compression format used for archiving, web downloads and for portable media devices such as mobile phones and iPods.

Non-linear editing

An approach to editing which does not depend on clips being assembled from the very start of the movie to the very end. Media can be added and removed from the sequence at any place and any time.

NTSC

National Television Standards Committee. The TV standard used in the USA, Canada, Japan and some territories of South America. NTSC standard DV footage has a frame size of 720×480 pixels and a frame rate of 29.97 frames per second.

OHCI

Open Host Controller Interface. A standard enabling wide-ranging compatibility for devices sporting FireWire or USB connections.

PAL

Phase Alternation Line. PAL is the TV standard for the United Kingdom, most of mainland Europe, Australia and New Zealand, as well as much of Asia. PAL DV footage has a resolution of 720×576 pixels and runs at 25 frames per second.

Progressive scan

Video recorded as full frames rather than interlaced fields. Each frame of progressive scan footage is of a higher quality than interlaced video shot on the same format, but motion smoothness is sacrificed in the process.

QuickTime

Apple's Video file format. As with AVI, the QuickTime MOV file extension can be applied to video files with many different types of compression. QuickTime is often used for preparing high-quality footage for archiving or compositing, making multimedia content for CD-ROMs or preparing progressive download footage for the internet.

RealMedia

Proprietary streaming media format developed and owned by RealNetworks.

Real-time editing

The ability to preview video or sound effects without the need for prior rendering.

Rendering

Processing video video or sound effects effects. Affected media is remade on the system's hard drive.

RGB

Red, green, blue – the colour channels used to compose many types of digital image.

Scene detection

Subdivision of video footage into smaller clips, based on points at which the camcorder was paused. Detection can be based on changes in the video's date/time information or sudden changes in on-screen content.

SECAM

The TV broadcast standard used in France. Note, however, that French DV equipment and DVD Video discs use the PAL standard.

Streaming video

Instant playback of internet video. Video is not stored on the viewer's computer and cannot be downloaded. The process requires specialized servers and can be costly.

S-video (Y/C)

S-video connections split video signals into two channels – luminance (Y) and chrominance (C). The S-video signal is of higher quality than composite, but of poorer quality than component.

Timecode

Timecode is the numerical labelling of every video frame. Timecode is typically ordered in terms of hours, minutes, seconds and frames.

VCR

Video cassette recorder.

INDEX

INDEX